Volcanoes and Wine

Volcanoes
& Wine

FROM POMPEII TO NAPA

Charles Frankel

THE UNIVERSITY OF CHICAGO PRESS

CHICAGO AND LONDON

The University of Chicago Press, Chicago 60637
The University of Chicago Press, Ltd., London
© 2019 by The University of Chicago
All rights reserved. No part of this book may be used or
reproduced in any manner whatsoever without written permission,
except in the case of brief quotations in critical articles and reviews.
For more information, contact the University of Chicago Press,
1427 E. 60th St., Chicago, IL 60637.
Published 2019
Printed in the United States of America

28 27 26 25 24 23 22 21 20 19   1 2 3 4 5

ISBN-13: 978-0-226-17722-9 (cloth)
ISBN-13: 978-0-226-60358-2 (e-book)
DOI: https://doi.org/10.7208/chicago/9780226603582.001.0001

Library of Congress Cataloging-in-Publication Data

Names: Frankel, Charles, author.
Title: Volcanoes and wine : from Pompeii to Napa /
Charles Frankel.
Description: Chicago ; London : The University of Chicago Press,
2019. | Includes bibliographical references and index.
Identifiers: LCCN 2019008606 | ISBN 9780226177229 (cloth :
alk. paper) | ISBN 9780226177366 (pbk. : alk. paper) | ISBN
9780226603582 (e-book)
Subjects: LCSH: Terroir. | Volcanic soils. | Wine districts. |
Volcanoes. | Wine and wine making.
Classification: LCC SB387.7 .F73 2019 | DDC 634.8—dc23
LC record available at https://lccn.loc.gov/2019008606

Contents

Italic indicates text boxes.

Preface · ix

1

VOLCANOES AND WINE

Volcanoes and Agriculture · 2
Volcano Types and Eruptions · 3
Volcanic Soil · 6
Volcanoes and Global Warming · 8
Volcanoes and Wine · 8

2

SANTORINI

The Dawn of Wine Making · 13
Sitting on a Time Bomb · 14
An Eruption of Mythic Proportions · 15
Renaissance and Vinsanto · 17
Santorini in the Twentieth Century · 19
Santorini Grapes · 20
Wine and Climate · 22
A Terroir Made of Pumice · 24
A Variety of Wines · 26
Guide Section: Visiting Santorini · 28

3

MOUNT VESUVIUS

The Bay of Naples · 37

The Vineyards of Pompeii · 38

The Fatal Eruption of AD 79 · 41

The Final Blow · 43

Pompeii's Burial Date Revisited · 45

A Vineyard Rises from the Ashes · 46

Vesuvius Grapes and Terroir · 47

Mount Vesuvius: Tectonic Setting and Magma Composition · 49

Italy's Orchard · 50

Lacryma Christi: Tears of Christ · 51

Living Dangerously · 54

Farming the Last Lava Flow · 56

A Look to the Future · 58

Guide Section: Visiting Mount Vesuvius · 60

4

MOUNT ETNA

Sicily's Garden of Eden · 67

Europe's Most Active Volcano · 68

Playing with Fire · 69

The 1991–1993 Eruption · 72

A Train Ride around Mount Etna · 73

Etna's Pistachios · 74

The Oranges of Mount Etna · 75

Strawberries and Wine · 76

A Brief History of Wine · 78

Etna Grape Varieties · 80

I Vigneri, Keepers of Tradition · 84

The Wines of Etna · 84

Etna's Fruit Brandies · 88

The Winemakers · 88

The Notion of Terroir · 91

Franchetti's Suite of "Terroir" Wines · 94

A Very Special Vineyard · 94

Guide Section: Visiting Mount Etna · 96

5

THE AEOLIAN ISLANDS

Malvasia: Nectar of the Gods · 108
Vulcano: Vines on a Time Bomb · 111
Lipari: The Central Island · 113
Stromboli: Fire and Wine · 114
Salina: The Hub of Malvasia · 116
Pantelleria · 117
Guide Section: Visiting the Aeolian Islands · 118

6

FRANCE'S HIDDEN VOLCANOES

Rift Zones in France · 124
A Volcano in Provence · 126
Vines Rooted in History · 127
Grapes of Auvergne · 130
The Comeback of Côtes-d'Auvergne · 130
Fire Meets Water: The Châteaugay Terroir · 131
The Hill of Corent · 134
Boudes, Chanturgue, and Madargue · 135
Wine and Pumice: The Neschers Terroir · 136
Guide Section: Visiting Auvergne · 137

7

THE CANARY ISLANDS

Vineyard History and Distribution · 145
Lanzarote and the 1730 Eruption · 146
Holes in the Ash · 147
Grape Varieties in the Canaries · 149
The Azores Islands · 150
Canary Wines · 151
Guide Section: Visiting the Canary Islands · 152

8

CALIFORNIA, OREGON, AND HAWAII

Napa Valley: A Tectonic Basin · 159

A Mosaic of Terroirs · 160

Oregon's Great Lava Fields · 162

A Mighty Flood · 164

Willamette Valley and Pinot Noir · 165

Ocean versus Lava: Pinot Noir Takes the Stand · 168

Hawaii and Coffee · 169

Wines of Hawaii · 171

Guide Section: Visiting California and Oregon · 172

Notes · 177

Bibliography · 185

List of Websites · 187

Credits · 191

General Index · 193

Index of Place Names: Geographical Names, Appellations, and Estates · 199

Index of Grape Cultivars · 207

Preface

Since the dawn of civilization, volcanoes have been praised for the wines produced on their slopes. In his *Natural History*, Pliny the Elder praises those of Mount Vesuvius—a reputation upheld today by the celebrated Lacryma Christi; in the Greek island of Santorini, sweet Vinsanto has been famous since the Renaissance; the best Malvasia wine comes from the volcanic islands of Salina and Pantelleria, off the coast of Sicily; and today the potential of Mount Etna's terroir is attracting winemakers from around the globe.

Other volcanic areas are less conspicuous, because they are older and worn down by erosion, like the little-known Côtes-d'Auvergne in central France; and in North America, many estates in California's Napa Valley and Oregon's Willamette Valley—famous for their Cabernet Sauvignon and Pinot Noir—are rooted in volcanic soil.

Why is there such a magic alliance between volcanoes and wine—and other crops, for that matter?

Many factors come into play that are described in this book: the chemistry and texture of the volcanic rock and soil—broken-up lava, pumice, and ash—and the volcanic landform itself, which influences weather patterns and provides a range of altitudes, slopes, and orientations, each blessed with its own microclimate, to fit a variety of crops. Not to mention the crucial role of the winemakers themselves, who both fear and cherish their volcano, endure its eruptions, and perpetuate farming and wine-making traditions based on centuries of experience.

I am a volcano lover myself and have taught volcanology both in France and in the United States. As I also happen to be a wine and terroir aficionado, having penned a couple of books on the topic, it was inevitable that I would at some point fuse my two favorite topics. This book thus aims to satisfy both readerships: fans of volcanoes, who will learn much about wine, and connoisseurs of good wine, who will learn a great deal about volcanoes.

With this goal in mind, each chapter begins by describing a featured volcano and its most famous eruptions, and continues by reviewing the history and present state of wine making on the volcano's slopes. Each chapter ends with a guidebook section that proposes geological, archeological, and wine-tasting itineraries. Indicative prices are listed for wine tasting and tours: they are based on 2019 figures, and are of course subject to change.

An introductory chapter first lays out a few basic notions of volcanology, so that the reader can become familiar with some of the lava types and eruption styles encountered in later chapters. It also presents the key aspects of a volcanic terroir: why the landscape and soil are so special.

I am grateful to University of Chicago Press and to my editor, Susan Bielstein, for being so supportive and patient while I penned this book, after I wrote a first version in French for my Paris publisher Dunod. And I wish to express my profound gratitude to all the winemakers and farmers who welcomed me on their estates and showed me the intricacies of their terroir, be it Stefanos Georgas and Paris Sigalas in Santorini; Antonio Dente and Vincenzo Oliviero on Mount Vesuvius; Salvo Foti and Frank Cornelissen on Mount Etna; Andrea Hauner in Salina; and Stéphane Bonjean in Auvergne, to name but a few: the reader will meet them throughout this book. I also wish to thank the wine "ambassadors" who helped me prepare my visits, as did, for instance, Sofia Perpera, Korinne Munson, and Stela Kasida for Santorini.

Most photographs in this book were taken during my field trips; many thanks to the professional and amateur photographers who provided the ones I lacked: credits are listed at the end of this volume, as well as a list of reference publications and websites for most estates mentioned in the text.

While *Volcanoes and Wine: From Pompeii to Napa* constitutes a virtual, introductory tour of volcanic estates, it is my hope that it will encourage the reader to travel in person to these wonderful sites and experience first-hand their cultures — in both senses of the word — and taste their wines.

Cheers from France!

Charles Frankel

CHAPTER I

Volcanoes and Wine

The notion of terroir—a sense of place—applies particularly well to volcanoes. These towering landforms have specific bedrock and soil, locally influence the climate, and offer a range of orientations and elevations to fit a multitude of crops.

The Earth is a showcase planet when it comes to volcanism. On any one day, approximately twenty volcanoes are erupting across the globe. The aggregate number of active volcanoes rises to sixty over the course of one year and totals six hundred across recorded history—that is, the past 2,500 years.

Our civilization is affected by volcanism in many ways, with some eruptions taking their toll of human lives, and others altering the climate on timescales of a few weeks or a few years. But the greatest consequence of volcanism might well be the creation of new landscapes, new minerals, and ultimately new soil. This is where plant life and agriculture fit into the picture.

Volcanic eruptions provide a range of chemical elements used by plants, both in solid and in gaseous form. Gases include the water vapor, carbon dioxide, and sulfur dioxide that volcanoes constantly expel into the atmosphere. Carbon and, to a lesser degree, sulfur are among the most important elements in the building blocks of life, such as amino acids and proteins. As for the elements distributed by eruptions in solid form, these include silicon, phosphorus, and a whole range of metals, such as calcium, sodium, potassium, iron, magnesium, aluminum, manganese, and other trace elements.

Locked inside lava flows and the fallout that rains from ash clouds, these solid elements need to be freed from their mineral cages in order for life-forms to absorb them — a breakdown process made possible by Earth's efficient water cycle, as well as by a vast chain of chemical and biochemical reactions.

Volcanoes and Agriculture

The role of liquid water in breaking down minerals and feeding plant life is one reason volcanoes are so important in agriculture. Their bulging masses create obstacles that deflect air currents. When a batch of humid air rises along the slope of a volcano, its temperature drops: water condenses and rains out onto the windward side of the obstacle, and dry air blows down the opposite lee side.

The windward, rainy side of a volcano is a haven for water-dependent plants, including a variety of crops like rice, fruits, and vegetables, but the dry lee side is also profitable for a whole range of crops that instead need little water and a maximum amount of sunshine, such as coffee, nuts, and grapes.

One prime example of this is just east of Naples, at Mount Vesuvius: Italy's most famous volcano. As described in its dedicated chapter, Vesuvius has long been the fruit basket of Italy, and despite increased competition from Europe's Common Market, the volcano still provides to this day the majority of the country's apricots, as well as its most prized tomatoes and cherries. The role of the volcanic landform is apparent when mapping precipitation and crop types: prevailing north winds focus rainfall on the northern side of the volcano, where most cherry trees are planted. The western sector gets moderate rainfall and enough sun to support apricot trees and tomato crops, whereas the dry southern and southeastern flanks — above Herculaneum and Pompeii — bear vineyards that need minimal water.

Mount Etna in Sicily is another good example of the role of orientation on a volcano, and of elevation as well: a sizable peak provides a wide range of altitudes, which translates into different climates at each level. Mount Etna thus possesses a low-elevation, hot agricultural belt for its prized oranges and citrus fruit, as well as an upper, cooler vineyard zone between 450 and 1,000 meters elevation (1,500–3,300 ft.): the cool nights, typical of the highest reaches, slow down the ripening cycle of the grapes and promote the formation of complex aromatic molecules, yielding quality wines that are now recognized worldwide. At the highest reaches of its northern flank, the volcano even harbors alpine birch and pine forests — an exceptional sight in Sicily — that were long harvested for timber and fuel wood.

Mount Etna's freestanding cone also provides a whole gamut of wind and rain exposure. Orange groves are located in the sunniest southwestern and southern sectors, whereas vineyards occupy the eastern half of the volcano, because morning sunlight is necessary to dry out any nighttime or dawn precipitation that might carry molds or other vine-threatening diseases. Even the northern sector of Mount Etna offers agricultural niches: in particular, the extra moisture carried by the northwestern prevailing winds benefit pistachio trees — a crop that reaches world-class excellence around the city of Bronte.

An interesting aside is that volcanoes can alter the regional climate when they erupt, which can significantly affect the characteristics of that year's vintage. A case in point is the May 1980 eruption of Mount St. Helens, in Washington State, which chilled the spring climate that year in the downwind Oregon estates to the point that the state's 1980 Pinot Noir won two gold medals in 1982 for the first time and was favorably reviewed in the *New York Times*.[1]

Volcano Types and Eruptions

Besides the range of orientations and altitudes they provide to crops, and their local effect on climate, volcanoes come in many different types and forms, and also display a diversity of eruption styles that dictate the type of substrate — rock texture and soil chemistry — available for agriculture.

Volcanism is the process by which a hot planet gets rid of its heat. A planet like the Earth is partially molten inside. There is a dense iron core at the center, with a solid inner part — due to extraordinary pressure — and a liquid outer part (with a temperature around 4,000°C, or 7,000°F). This molten iron never reaches the surface and does not take part in volcanism. Above the iron, starting at a depth of approximately

3,000 kilometers (1,800 mi.), and stretching almost up to the surface, is a mineral paste known as the mantle. It also contains iron, but mostly silicon and magnesium, as well as calcium and aluminum, and other metals in small amounts, all bound into interlocked crystals by a great deal of oxygen. There are also minute amounts of volatile molecules dissolved in the mantle, principally water, carbon, and sulfur oxides. The temperature ranges from 4,000°C at the bottom of the mantle to about 1,000°C at the top (7,000°F–1,700°F), hot enough for the mineral paste to flow and circle in great loops, like molasses on a stove, but at a very slow rate: a couple of centimeters (about an inch) per year.

Above the churning mantle lies a lid of cooled, rigid "scum": the Earth's crust. Only 5 to 50 kilometers thick (3 to 30 mi.), depending on the location, the crust is the result of countless volcanic eruptions that tapped the upper mantle over billions of years to coat the surface of the planet with flows of molten rock that chill into place: the icing on the cake.

Volcanism is therefore the process that moves fluid mantle material to the surface. It does not occur everywhere: the mantle is very hot but rarely hot enough to melt and send streams of liquid rock upward. Exceptional places where the mantle is hotter than average and where volcanism does occur are known as hot spots. Dilated by the extra heat, the mineral paste ascends toward the surface, in the same way that a hot-air balloon rises through cooler air. Moreover, as pressure declines on the way to the surface, the hot, buoyant rock begins to melt. This is similar to taking a kettle of hot water up a mountain: the pressure drop causes the water to boil.

As the rising blob of hot rock impinges on the Earth's crust, it causes it to bulge and fracture: the melt then rushes through the fissures and erupts at the surface. At first, the molten rock, known as magma, can contain a lot of dissolved gas that makes it fizz, like a soda bottle opened for the first time: the magma sprays skyward in the form of a lava fountain, also known as a Hawaiian eruption, in reference to the most famous hot spot on Earth that created the Hawaiian Islands. After most gas is flushed out, the magma simply oozes out of its fissure or crater and proceeds downslope as a peaceful lava flow.

In hot-spot settings, minerals come from great depth in the mantle and have a metal-rich chemistry that makes the resulting lava particularly fluid: flows cover great distances and build shallow-sloped shield volcanoes. In view of the large volumes of magma pumped up by hot spots, such shields can reach impressive sizes and elevations. Hawaii's Mauna Loa volcano holds the record of the largest volcano on Earth, with an estimated volume of 75,000 cubic kilometers (18,000 cu. mi.) and an elevation of 9,170 meters (30,085 ft.) above the seafloor.

Another setting for volcanism on Earth is provided by shallow convection loops of the upper mantle that act like the rollers of a conveyor belt and tear the crust apart into great segments, or plates, that move relative to each other — this is the slow ballet of plate tectonics. Where two plates move apart, magma rises at the seams and pastes new volcanic crust along their boundaries: such locations are known as rift zones and often develop in oceanic settings, deep under water (mid-ocean ridges). Hence, they are rarely associated with agriculture, or with wine making for that matter.

Other plate margins have not experienced a hot-rock pasting for a while, and they have cooled, contracted, and densified to the point that one plate flexes downward (creating a topographic trench) to slide under its neighbor and sink back into the mantle: a process known as subduction. During this process, the sinking slab heats up and expels water vapor and other fluids that then work their way back up through the overlying hot mantle. Because the injection of fluid into hot rock is one mechanism that can cause rock to melt, this often generates magma, which ascends to build a chain of volcanoes behind the subduction trench. Examples include the Cascades range in the states of Washington and Oregon; the volcanoes of Central America and of the South American Andes; Mount Vesuvius and other volcanic landforms along the coast of Italy; and Indonesia, the Philippines, and Japan, along the Pacific Ring of Fire.

Subduction volcanoes often host explosive eruptions because of the quantity of volatiles — principally water vapor — that end up in the erupting magma. Water also affects the type of magma generated, promoting a high proportion of silica in the brew and making it particularly viscous. The combined effect of high volatile content and viscosity blows the magma apart as it rushes up the volcano's chimney: jets of fragmented magma — particles named *pyroclasts* by geologists — billow skyward to form dense clouds, or plumes, that drop their content downwind of the craters. Such outbursts are Plinian eruptions, in reference to the famous eruption of Mount Vesuvius in AD 79 described by Pliny the Younger.

The fallout particles, which are often ridden with holes by the escaping gases, bear different names depending on their porosity — very porous ones are *pumice* — and on their size, for example, *ash* when they are fine, *lapilli* or *scoriae* when they are the size of a nut, and *blocks* or *bombs* when they are the size of a fist or larger.

Finally, the fallout can take on a catastrophic form when the pyroclasts expelled by the eruption build up such a dense column in the atmosphere above the crater that it collapses under its own weight and rolls over the landscape like an airborne tsunami: such pyroclastic flows can level and blanket entire towns, as they did Pompeii and Herculaneum in AD 79, and more recently Saint-Pierre in Martinique in 1902.

Volcanic Soil

With so many different geological settings and eruption styles, it is no wonder volcanic terrain comes in a variety of textures and profiles, adapted to various degrees to agriculture and vine growing. Some lava flows are hard and unbroken and require working over with crowbars and bulldozers, or centuries of erosion, to become farmable. Pumice and ash fallen from the air, however, are readily tillable. Both often occur together on the same site: trenches and cross-sections through volcanic fields frequently show piles of lava flows interspersed with layers of ash fall.

Besides ground texture, the chemical makeup of lava and ash naturally affects agriculture. Some elements brought up from the mantle by volcanic eruptions are particularly valuable to plant life and enrich the soil. Boron, for example, plays a fundamental role in cell division and influences flowering and fruit set, thus directly affecting crop yield. Potassium is extremely important in regulating sugar and acid content — a role it plays in grape juice and wine in particular.

On somewhat aged soils, on which crops have been grown and rotated over centuries, these elements have already been soaked up by countless plants and removed from the land. For new crops to receive their proper share, the soil needs to be spiked anew with mineral nutrients, and repetitive eruptions are a great way to do so. Compared to lava flows, which might take decades or centuries to break down sufficiently to provide these precious elements to crops, explosive, gas-rich eruptions are particularly efficient in doing the job. The material not only is blown to bits by the expanding gases but also can be sprayed far and wide, covering large areas. If the output is diluted, it can shower existing crops and reach the soil without harming the vegetation. Too thick an ash cover, however, can choke the plants, as is often the case on the flanks of Indonesian volcanoes.

The big advantage of ash fall is that its minute particles offer a greater surface-to-volume ratio than large blocks, facilitating their interaction with water, bacteria, fungi, and other agents capable of breaking down the volcanic minerals to free up their elements. Leaching by rain and run-off water works best if the water is slightly acidic, as is often the case on volcanoes: erupting plumes provide sulfur that combines with water to form sulfuric acid and with carbon to form carbonic acid.

The process is all the faster if the starting material is poorly crystalline and glass-rich, which is typical of airborne ash. Once the ash is mixed with organic-rich humus already present on the ground — from the decomposition of previous plant life — the result is a rich volcanic soil. Volcanic soil, known as Andisol, constitutes one of the thirty or so major soil groups recognized by geologists.[2]

Andisols are usually fluffy and of low density, and thus easy to plow. There are even cases of volcanic soil being shifted from place to place to support agriculture, such as in the Canary Islands, where volcanic cones are mined for the material, which is then loaded on trucks and spread in the lowlands to grow a variety of crops, like beans and potatoes.

Volcanic soils also have drawbacks, however. They are easy to disturb, can be washed away by floods, and can even fail in dramatic landslides, as happened at Sarno, near Naples, where in 1998 a slope failure claimed 161 lives. Mudflows can also rush down gullies in the wake of water-rich eruptions and run over entire towns, pooling and setting like concrete.

However, by and large, volcanic soil is a precious, fertile substrate for agriculture. Andisols support coffee and tea crops, maize, rice and potatoes, tobacco, and fruits and vegetables as a whole. Grapevines and wine, the main topic of this book, are rarely cited in that list, though, and for good reason: grapes are mostly grown outside tropical regions, where most active volcanoes are located. In the wine belts of temperate regions, the substrate is most often limestone and sandstone, granite and schist, gravel and sand, with little soil. But when there is a match in these temperate zones between volcanoes and vineyards — the examples explored in this book — the result is often spectacular.

VOLCANOES AND GLOBAL WARMING

Global warming is a fast-paced, troublesome evolution of the worldwide climate that has a direct impact on agriculture. The rise in temperature and the change in precipitation patterns are so rapid that farmers and vine growers need to plan medium- and even short-term adjustments for their crops.

With coffee as an example, rising temperature and precipitation changes are directly threatening plantations for two reasons. A hotter environment is detrimental to the health of the plant and the quality of its coffee beans. Also, many pests and fungi thrive at higher temperatures and follow the temperature rise upslope to infest the crops. In Tanzania, for instance, the nasty coffeeberry borer *H. hampei* has climbed 300 meters (1,000 ft.) in ten years.

Farmers are faced with two solutions: replacing their crop with other ones best suited to the new temperature, or moving crops upslope to provide them with the cooler temperatures they require. Clearly, the latter solution is impossible in the plains or on a plateau, where relocating crops higher up is impossible. Volcanoes, however, play an interesting role, in that they allow for such a vertical migration.

Most estimates predict a rise of global temperatures on the order of 2°C (3.6°F) by the year 2050. To remain at a given temperature, crops will need to move upslope approximately 300 meters (1,000 ft.) in the same time frame. This might not be possible on the Ethiopian Plateau, where coffee crops are already at the highest altitude and have nowhere to go. The situation is brighter on a large and tall volcano like Hawaii's Mauna Loa, where the Kona coffee belt can creep upward into what is presently a mix of ranch land and jungle that can be claimed for farmland.

Vineyards are another example; this situation is best illustrated by Mount Etna in Sicily. Above the wine estates, most of the volcano's upper slopes are uncultivated, except for occasional chestnut groves, so that vines can progressively be planted higher up on bare lava flows. Indeed, the shift is already occurring: winemakers are conscious of the better quality of wine produced at higher elevations, regardless of global warming. Hence, the shift upslope in the future is unlikely to be considered a negative constraint. It does mean purchasing new land, however, so that small family estates that cannot purchase new lots might suffer to some extent — or shift to other grape cultivars adapted to higher temperatures. Another disadvantage is that moving upslope on a volcano means more exposure to lava flows that descend from the summit: this means a higher risk of losing land, but it is a risk that populations living on volcanoes are willing to accept.

Volcanoes and Wine

Volcanic soil provides abundant nutrients for plant life, and like most plants, grapevines appreciate the fertility it provides. For the vines to yield aromatic grapes and good wine, however, that fertility needs to be tempered. Vine growers are conscious that the plant must be somewhat deprived of soil and water. Many farmers periodically slice up the ground around the vine stocks to cut off superficial roots that could feed on the fertile topsoil, and thus encourage the propagation of roots vertically into the bedrock, where the plant is encouraged to seek its moisture and mineral elements.

On many volcano slopes, this pruning is unnecessary, because broken-down, fertile soil is still lacking: such is the case on the Greek island of Santorini, where vines stick out directly from a blanket of coarse, white pumice. The situation is nearly the same on Vesuvius and Mount Etna, where mature soil has developed in only a few places at the surface. Elsewhere, as in Oregon's Willamette Valley, the basalt flows are so old (more than fifteen million years) that deep soil has developed, as described in the chapter devoted to the American Northwest.

Besides their coarse texture and limited fertility, scoria and pumice present another advantage: their holes, or vesicles, created by the blowout of gases from the magma, offer a spongelike texture that traps and concentrates moisture in dry environments, then distills the little water needed for the vines to produce quality grapes.

Scoriae and rocks at the surface also soak up solar heat during the day, then return it to the vines as infrared rays after sundown, acting as miniature radiators that lengthen the daily ripening cycle of the grapes.

An ongoing object of debate is the influence of chemical elements in the ground on the flavor and quality of a wine: what is often mistakenly stated as the principal "terroir effect." With exceptions for the roles of calcium and potassium in the acidity of the grape juice (and therefore of the wine); of sulfur, which is of course plentiful on volcanoes; and of iodine, provided by sea spray in coastal settings, many geologists and oenologists are not convinced that the purely chemical dimension is so crucial.

In conclusion, it must be stressed that terroir — the sense of place — is a combination of many parameters, chemistry being only one of them. The texture of the soil, along with its percentages of stone and rock cover, is crucial. Even color is important, as the red wavelengths of light, reflected from red-tinted rocks and soil, promote the formation of anthocyanin molecules in wine, which govern its color. And on a wider scale, slope, orientation, and of course microclimate are essential as well.

The choice of the grape cultivar itself is part of terroir, as winemakers have experimented over the centuries with selecting and crossing grape stock to create cultivars best adapted to their land. On volcanoes have emerged very specific grape varieties — Piedirosso on Mount Vesuvius,

Assyrtiko on Santorini, Nerello Mascalese on Mount Etna—that give their best results on their native volcano, and less spectacular results when transplanted elsewhere.

Developing specific cultivars is only one way in which winemakers enter the big picture of terroir—not only a sense of place but also a sense of time and tradition over the centuries. It is this complex and historical interplay that comes to life in the following tour of the world's greatest wine-producing volcanoes.

CHAPTER 2

Santorini

The Greek Island of Santorini suffered one of the most powerful eruptions in the history of mankind, which devastated the burgeoning nations of the Bronze Age in the Mediterranean basin. Today, the white pumice is peppered with vine stock—the very same varieties that were grown more than three thousand years ago by the Minoan civilization. Santorini wines range from dry whites to the famous, amber-colored Vinsanto.

Santorini Island, in the Aegean Sea, claims to be the first wine-producing volcano in history. During the Bronze Age, Minoan ships delivered its wine as far as Egypt, and during the Renaissance, Venetian merchants shipped its celebrated Vinsanto to churches and cathedrals across Europe.

Eight hours from Athens by boat, three to four hours from most European capitals by plane, Santorini is a famous holiday destination, on account of its extraordinary setting and dramatic history. Cruise ships and private yachts drop anchor in the middle of the horseshoe-shaped caldera, below the towns of Fira and Oia perched on the cliffs above. Every morning, tourists swarm the narrow streets, visit the Orthodox churches, jewelry shops, and Greek taverns, and by day's end, return to their ships and sail off into the sunset.

For those who are really intent on visiting the island, there is another side to Santorini: the backside of the volcano, with authentic villages off the beaten track and vineyards running down to the sea. In the 1890s, the ruins of a Minoan city were unearthed on the southern shore of the island, shedding light on one of the greatest civilizations of antiquity. The

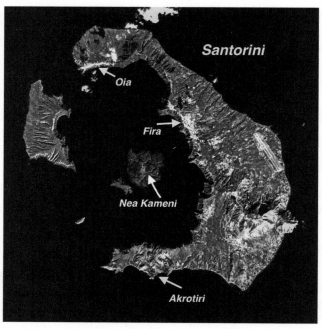

Location (top) and satellite view (bottom) of Santorini Island, Greece.

Minoans ruled the Mediterranean from 2000 to 1500 BC, then mysteriously collapsed.

The downfall of Minoan civilization is blamed in part on the Santorini eruption of 1620 BC, which obliterated the island and generated tidal waves that devastated Crete's northern shore, wrecking the Minoans' harbors and fleet. Plato's mythical account of the submerged city of Atlantis is thought by many to reflect this catastrophic sequence of events.

Abandoned for centuries, Santorini was eventually resettled by Phoenicians and Mycenaeans, who restored the olive groves and vineyards. The island's indigenous grapes, in particular, met with renewed success. They were sun dried to yield a sweet white wine, known as Vinsanto, that became famous during the Renaissance, when Venetian merchants shipped it across Europe. It became the sacramental wine of Catholic and Orthodox churches, spread to the nobility and bourgeoisie, and to this day remains the most famous wine from Santorini, although today the island's dry whites, pressed from the same Assyrtiko grape, also rank among the very best Greek wines.

Over two dozen other grape varieties are grown on Santorini, albeit in minor proportions compared to the ruling Assyrtiko. All are ungrafted, original varieties, as the phylloxera blight that destroyed most vineyards in continental Europe, at the close of the nineteenth century, was unable to thrive in the island's dry pumice. Hence, Santorini can claim the oldest line of ancestry of any wine known to man, down through five millennia of history.

The Dawn of Wine Making

The training and selection of wild vines, and the mastering of fermentation and wine-making techniques, first began in the fertile valleys of Mesopotamia. Grape seeds together with pine resin — one of the first wine preservatives — were found at an archeological site in Iran, some seven thousand years old, setting the record for the oldest evidence of wine making discovered so far.

When ancient Egypt replaced Babylon as the leading civilization of the Western world, it too turned to wine making, for domestic consumption and for trade, but it was soon challenged by the blooming civilization based in Crete that shipped its wine across the Mediterranean.

Archeologists date the rise of this powerful Bronze Age civilization to approximately 2500 BC. It flourished for ten centuries and reached its apex around 1700 BC, with the construction of sophisticated palaces in Crete, most famously at Knossos. Frescoes depicting bull-riding gymnasts are believed to have inspired the myth of Theseus and the Minotaur, so archeologists named this towering culture the Minoan civilization.

About 150 kilometers (roughly 100 mi.) north of Crete, Santorini island hosted a prosperous Minoan colony. It produced not only wine but also, according to inscriptions engraved on amphorae, a whole range of different wines: at least nine varieties for the last year on record, when life on Santorini came to a dramatic end. In 1620 BC or thereabouts, according to carbon 14 dating, the volcano central to the island unleashed its fury in one of the most powerful eruptions known to mankind.

Sitting on a Time Bomb

One doesn't need to be an expert to figure out that the island of Santorini was rocked by a cataclysmic eruption: nearly the entire landscape is covered by a blanket of cream-colored pumice, up to 60 meters (200 ft.) thick in places.

Thirty centuries later, quarries were dug into the pumice: a precious construction material in view of its low density, insulation properties, and good compressive strength. The location of the quarries next to the waterfront made it easy to load and ship off the material to domestic and foreign markets. While they were quarrying Santorini's southern shore in the 1860s—the pumice was shipped to Egypt for the construction of the Suez Canal—workers unearthed the walls of a city that had laid buried for three thousand years, bringing back to light the marvels of the Minoan civilization. The ruins are named Akrotiri, after the village nearby.

During the Bronze Age, Santorini was a dormant volcano, and the Minoans were probably not aware that they were sitting on a time bomb. Geologists believe that a large volcanic cone rose in the center of the sea-filled caldera. Molten rock silently accumulated in the underground magma chamber, beneath the seafloor, while gases fizzed out of solution to put the whole system under pressure.

Magma motion and gas circulation underground can trigger earthquakes, and there is no doubt that tremors repeatedly rocked the island. In fact, there is evidence that a major earthquake occurred months or weeks before the cataclysmic eruption. In the Minoan city of Akrotiri, the walls of several houses collapsed, others are slightly tilted, and one flight of stone steps, which visitors can view in the central part of the ruins (the so-called Delta complex), is split straight down the center by the violence of the jolt, estimated to have reached at least a 7 on the Richter scale.

At that point, villagers probably fled the island temporarily. But many returned to clear the rubble and repair the damage, as attested by heaps of loose stones piled up in the streets and buckets of mortar and stone hammers lying about.

Soon enough, the volcano resumed its dramatic reawakening. With the ominous rumbling and a new shower of ash falling on the city, did the remaining population have time enough to flee? Most probably — unlike Pompeii, not a single human skeleton has been found in the ruins so far, or any valuables such as gold, indicating that the retreat may have been somewhat organized and timely.

An Eruption of Mythic Proportions

The eruption has been dated to around 1620 BC — give or take ten or twenty years — by several independent techniques, including carbon 14 dating of an olive tree uncovered in the ruins of Akrotiri. From the sequence of stacked layers of ash, volcanologists believe the eruption began with a series of explosions, as gas under pressure began to clear the chimney of the volcano. The rising magma mixed with seawater, and steam explosions then blew to bits the plug of hard rock clogging the volcano's chimney.

The result was a sudden drop in the confining pressure inside the volcano and the rapid expansion of gas bubbles in the magma, violently propelling the pumice froth skyward, in the same way that uncorking a bottle of champagne causes it to spray. A thick jet of pumice roared out of the widening crater, building up an ash column that rose an estimated 30 kilometers (20 mi.) into the stratosphere, according to the layout and thickness of ash deposited downwind. A huge cloud eclipsed the sun, moving across the eastern Mediterranean and into the Middle East.

After expelling the upper batch of magma and considerably widening its crater, the spewing volcano drained an even larger amount of seawater into its bowels, which cascaded onto the incandescent rock and flashed into steam, blowing up what was left of the volcano and its magma chamber into a cloud of hot pumice that spread in all directions, like the ground-hugging surge of a nuclear blast.

The shock wave toppled all obstacles in its way, including entire wall sections in Akrotiri that collapsed southward. The base surge dumped 10 meters (30 ft.) of hot ash on the city and countryside, and spread out over the sea, catching up with and engulfing any ship that had not left early enough to avoid the volcano's fury.

On the site of the gutted central volcano, the eruption was not over yet. Seawater swirled back into the widening blister and set off a second round of base surges, propelling an even larger and much wetter ring of hot pumice up in all directions, blanketing Santorini under 50 more meters (165 ft.) of ash — about seven times the thickness that covered Pompeii when Mount Vesuvius blew its top.

What was left of the hollowed-out magma chamber collapsed under its own weight, blowing out the last dregs of magma through peripheral ring faults, in a round of detonations probably heard well over 1,000 kilometers (620 mi.) downwind, as far as Egypt, Iran, and Iraq.

The total volume of erupted material reached an estimated 60 cubic kilometers (14 cu. mi.), which ranks as one of the most voluminous eruptions of the past ten thousand years, on par with the Tambora eruption (50 km³, or 12 cu. mi.) of AD 1815.

There is no doubt that such a cataclysmic eruption left civilizations in shock across the Mediterranean. Tsunamis from the collapsed caldera spread out from ground zero, unleashing a huge amount of energy onto Crete's northern shore and devastating Minoan harbors and palaces. They probably retained enough momentum to strike the shores of Egypt and Palestine with waves over a dozen meters (40 ft.) in height, equivalent to the tsunami that devastated Indonesia in 2004.

The drawback and then surge of coastal waters might well have formed the basis for the story of Moses parting the waters of the Red Sea and engulfing Pharaoh's army. As for the cloud of ash that drifted across the Mediterranean into the Middle East, blotting the sun, generating blood-red sunsets, and ruining crops under 15 centimeters (0.5 ft.) of fallout, it might have provided inspiration for the plagues of Egypt as well.

It was once believed that the disaster caused the demise of the Minoan civilization, a hypothesis championed by Spiridon Marinaros, the Greek archeologist who led the Akrotiri digs. Having lost their Santorini colony, their fleet devastated by tidal waves, the Minoan people felt abandoned by their gods and were unable to recover.

A critical review of the timing of events no longer supports such a straightforward link. Initially, the eruption was believed to have occurred in the 1500s BC, coinciding with the final collapse of the Minoan civilization, but as dating techniques improved, the age of the eruption slipped back to 1620 BC. Although it now precedes the end of the Minoan era by a full century, the eruption might still correspond to an earlier crisis that occurred during the Neopalatial Period in Crete's history, when many palaces were destroyed and needed rebuilding.

Whatever role it played in the downfall of the Minoan civilization, the eruption completely buried the colony of Santorini. The smoldering island remained uninhabited for over two centuries, blanketed by dozens of meters of hot pumice, until the cooled ash began to nurture seeds blown in from the spared hilltops, and vines and olive trees reclaimed the island.

Renaissance and Vinsanto

Eventually, civilization did return to Santorini. The Greek historian Herodotus (c. 484–c. 425 BC) mentions a Phoenician colony on the island around 1300 BC. In written records, the island goes by several names: first Strongyle ("the round one"), then Kalliste ("the beautiful"), and finally Thira, the name of a Dorian warlord who ruled over the island. Under the reign of Alexander the Great (356–323 BC), Thira served as a stronghold for the Greek fleet in its drive to control the Aegean Sea and the eastern Mediterranean.

The island lost much of its strategic importance during the Roman and Byzantine eras but rose to prominence again in the thirteenth century, during the Crusades, when European armies repeatedly crossed the Mediterranean to reach the Holy Land, using the volcanic island as a waypoint to stock their ships with wine and fresh produce. Merchants from Venice realized the commercial potential of the Greek Islands and set up a duchy in Naxos, with Thira as a southern annex. Christian Orthodox churches were well established by then, and the island was baptized Santa Irini, after Saint Irene, one of its patron saints — the name later became Santorini by contraction.

Cotton and wine were the two major crops developed on the island by Venetian merchants. Wine in particular benefited from centuries of careful crossing and selection of local grape varieties, many going back to Phoenician and even Minoan times.

Assyrtiko is the leading white grape on Santorini. Today it accounts for over 75 percent of the island's vine stock — suggesting that it is the oldest variety — followed by Athiri and Aidani. All three are used in the production of Vinsanto,[1] the wine that Venetian merchants sold across the Mediterranean. The wine owes its originality and quality both to its grapes and to the age-old wine-making process. Grapes harvested in July are sun dried out in the open for two weeks, which evaporates much of their water and concentrates sugars and aromatic molecules. Maceration often extends until Christmas, and the wine is then aged in oak barrels for a minimum of two years. During this time it acquires its characteristic amber color and rich bouquet of orange peel, dried apricot, figs, and nuts.

Besides its rich aromas and unctuous mouthfeel, Vinsanto conquered Europe because its high alcohol and sugar content helped it endure long shipping journeys. Traveling well beyond Italy, it became the official eucharistic wine of the Russian Orthodox Church. The wine's reputation and economic value were such that when the Turks took over Santorini from 1579 to 1821, they maintained its production, despite the Muslim ban on alcohol. Commercial ties were strengthened with Constantinople (now Istanbul), Russia's port of Odessa, and Alexandria in Egypt. By the mid-nineteenth century, even France was importing Vinsanto.

Then came the war of independence that freed Greece from Turkish control. Santorini played a major role in the war, on account of its strategic location and navy power, and it joined the new Greek Republic in 1830. The wine trade continued unabated through these times of turmoil. Even the phylloxera louse, which destroyed most European vineyards in the late 1800s, did little to slow Santorini's wine production: the insect was unable to colonize the island's powdery ash, which collapsed its burrows and choked its larvae. As a result, most Santorini vines have original rootstock that never needed to be replaced.

Santorini in the Twentieth Century

Wine making flourished on Santorini throughout the twentieth century. Besides the Vinsanto sweet wine, dry whites, reds, and rosés were also produced. The cotton and textile industry contributed to the island's economic growth, as did vegetable crops, especially tomato farms. About a dozen tomato canneries were built on Santorini: their brick chimneys still stand today as a testimony to the island's golden age of farming.

Two disquieting events reminded the islanders, though, that prosperity could once again be toppled by the volcano's restless mood: a small eruption in 1950 and an earthquake in 1956.

Santorini belongs to a special class of volcanoes known as resurgent calderas, which undergo cataclysmic eruptions in a cyclic fashion. After a blowout, such as the one in 1620 BC, molten rock slowly refills the underground magma chamber. At Santorini, the interval needed to replenish the magma chamber appears to be about twenty thousand years, which theoretically grants the islanders today a respite of fifteen millennia or so.

During the replenishing process, trickles of lava spill onto the seafloor and build up an underwater bulge. The mound surfaced over two thousand years ago, to form the twin Kameni islands (*kameni* means "burnt" in Greek). On Nea Kameni — the "newly burnt," more recent island of the pair — the last eruptions were recorded in AD 1570, 1866, 1925, and most recently in 1950, when lava fountains and billows of ash made for a spectacular, albeit harmless show when viewed from the caldera cliff 2 kilometers (about 1 mi.) away.

Geological activity in the area can also take the form of earthquakes. On July 9, 1956, an earthquake shook the seafloor halfway between Santorini and Amorgos Island to the northeast, registering 7.6 on the Richter scale and bringing down many houses. The quake was responsible for at least fifty casualties on the island and sparked an exodus toward the mainland, as farmers abandoned their orchards and vineyards.

During the decades that followed, the rise of tourism brought a second blow to the island's agriculture, offering easier, more lucrative jobs than the backbreaking tending of ground-hugging vines. Santorini's vineyards would have collapsed further had it not been for the support of the island's cooperative, Santo Wines, which encouraged their upkeep. Another boost came from the official recognition of Santorini wine, awarded an OPAP appellation in 1971 (OPAP stands for *Appellation of Origin of Superior Quality*, a ranking that was later upgraded to PDO — *Protected Designation of Origin* — under the European classification scheme).

The exodus of farmers and the rise of tourism left their mark. Deserted terraces on hillslopes, underscored by dark lava walls, no longer support any crops. The vineyards have now shrunk to the point that they occupy only the easiest, most accessible locations, especially because winemakers realize that the hard-to-work hills no longer provide a clear advantage over the sunny, well-ventilated lowlands.

The vines are low-lying, scattered across the landscape and seldom in neat rows, so that they are easily mistaken for wild bushes. Santo Wines, the island-based cooperative, supports small landowners by setting an attractive floor price to buy up their grapes, encouraging them to pursue their traditional practice. The cooperative purchases 60 percent of total grape production on the island; independent estates account for the remaining 40 percent, with 2,500 landowners tending to 1,200 hectares (3,000 acres), which amounts to about half a hectare (about an acre) per owner.

Santorini Grapes

Santorini hosts close to thirty grape varieties: most are indigenous to the island, selected and perfected in situ over thousands of years. Others were imported in more recent history from neighboring islands or the mainland. White varieties are by far the most popular, led by the local star: Assyrtiko.

Assyrtiko

The origin of this grape probably dates back to the Minoan era, over 3,500 years ago. Assyrtiko is characterized by small round grapes, with a juice that remains acidic late into the growing season — a welcome quality, because acidity lends a tangy freshness to the wine, as well as good aging potential. Considering that Assyrtiko also expresses the subtleties of its terroir in its palette of aromas, the variety is now in high demand all over Greece. Winemakers are planting it in Macedonia and Attica, where it develops less of an acidic bite and takes on a softer, fruitier character. On the volcanic pumice typical of Santorini, Yiannis Paraskevopoulos, cofounder of the Gaia estate, describes it as a grape "with claws," one that yields wine with a strong character.

Athiri

Athiri is the ideal partner for Assyrtiko in white wine blends. It is also an indigenous grape: the name *athiri* is believed to mean "coming from Thira," although it could also stem from *thirio*, meaning "wild beast." Its grape has a thinner skin than Assyrtiko and yields a wine with less acidity and alcohol, and more of a fruity flavor. Athiri is also grown on the mainland, in Macedonia and Attica, but its greatest success story is in the island of Rhodes, where it yields a popular sparkling wine.

Aidani

Aidani is Santorini's third white variety. Like Athiri, its acidity and alcohol content are on the low side. With a distinctive floral bouquet, Aidani brings character to white blends and can also stand alone as a varietal wine: it is produced as such by the Argyros and Sigalas estates.

Mavrotragano

Red grapes are somewhat of an oddity on Santorini, accounting for only 5 percent of the vineyard, but they are experiencing a comeback, thanks to dedicated winemakers like Paris Sigalas and the late Haridimos Hatzidakis, who placed their trust in the Mavrotragano variety (2 percent of the vineyard). Once on the brim of extinction, the grape is native and exclusive to Santorini. Its name means "black and crunchy," on account of its small, thick-skinned grapes that provide abundant tannin as well as a dark color to the wine. Initially vinified as a mellow dessert wine, Mavrotragano has the potential to penetrate the market as a dry red, capitalizing on its jammy aromas and earthy notes that bring to mind the Nebbiolo wines of Italy.

Mandilaria

Mandilaria might not be indigenous to the island, as it is found in Rhodes and Crete as well, where it yields a popular wine. On Santorini, its alcohol content is marginally low, but the grape makes up for this deficiency with a thick skin and abundant tannin. The wine exhales a bouquet of red berries and licorice, and is a minor component of red blends on the island.

Voudomato

Indigenous and exclusive to Santorini, Voudomato — which means "bull's eye" in Greek — releases a floral bouquet and either is blended in dry reds, as a partner to Mavrotragano, or stands alone as a mellow dessert red (Hatzidakis estate).

Besides these three white and three red grapes, Santorini hosts another twenty varieties, albeit in very small quantities. To protect indigenous grapes, the Santo Wines cooperative has established a conservatory and bans foreign grapes from the island.

Wine and Climate

Santorini varieties owe their character to their special environment, in terms of both climate and soil, and to the stewardship of vine growers who devised original pruning techniques to take full advantage of the conditions: a combination that gives full meaning to the word *terroir*.

Santorini's climate is not naturally favorable to agriculture. Isolated in the middle of the Mediterranean, within reach of the dry, hot winds that blow from Africa in the spring, Santorini is technically a desert, with 3,600 hours of sunshine a year and a mere 200 millimeters (8 in.) of rainfall, mostly during winter. The islanders store the little rainfall that they manage to collect in cisterns on their flat rooftops, draw water from a few scattered wells, and the balance is provided by tankers.

Wind is the other major agent of Santorini's climate. During winter and spring, southern winds blowing from Africa — known as *ostria* — can reach damaging speeds. During summer and fall, the northern *meltemi* winds are more stable and less damaging.

The wind's main impact is a sandblasting effect: it lifts the fine volcanic ash and whips it against the vines. Most of the damage occurs in the spring, when the plant's "eyes"—the fragile buds that will later turn into grape-bearing branches—are exposed.

Vine growers developed a clever pruning and training technique to circumvent the problem; this consists of flexing the main branch of the vine, close to the ground, into a circle. After several years of training, the vine is woven several times around into the form of a crown or basket. Every spring, the vine grower selects which eyes around the crown—usually four or five—should be promoted to grape-bearing branches that year. The buds are usually chosen on the inner periphery of the crown, protected from the outside winds and their sandblasting effect. Named *kouloura*, the basket form also provides shading to the grape clusters and limits the leaves' evapotranspiration, saving the little moisture the plant manages to collect.

One Santorini phenomenon in particular provides water to the grapes. The caldera acts very much like a cauldron.[2] During the day, seawater evaporates, and the vapor rises along the cliff walls. When the temperature drops at night, mist condenses and rolls down the outer slope over the vineyards, blessing them with a precious ration of water. The *kouloura* acts like a mist trap for the water, which is further absorbed by the porous soil.

In other climes, grapes shrouded in mist would most likely rot. On Santorini, the dry winds that blow over the vineyard stop the process by midmorning, acting in this case as a protective rather than destructive agent. In this healthy environment, vine stock can reach impressive ages. With no enemies to fear—as already mentioned, even the phylloxera louse was unable to dig itself a stronghold on the island because of the dry, granular nature of the soil, devoid of soft clay—Santorini vines grow very old. After one hundred years or so, it is customary to cut off the vine at ground level and start weaving a new basket from the same rootstock. The operation can be repeated every century, four or five times over, so that many vines on Santorini are in fact four hundred or five hundred years old, dating back to the Renaissance.

Some vines go back even further. To obtain a new plant, vine growers can bend a cane from an old vine into the ground next to it, where it takes root: this is the layering or marcottage technique. When the new offshoot is mature enough, after four or five years' growth, the "umbilical cord" connecting it to its parent vine is cut off, and a new, independent plant is born. It is technically a clone, and by this method the same individual stock can be regenerated indefinitely. A number of vines still bearing fruit could in essence date back to Greek or Phoenician times, 1000 BC, or even to Minoan times — some wild vines could have survived the 1620 BC eruption and been brought back into production by the returning settlers. Uncorking a bottle of Santorini Assyrtiko is like opening a time capsule that takes us back to the very source of Western civilization.

A Terroir Made of Pumice

Santorini's terroir is deemed exceptional, not only on account of its micro-climate and clever vine-training techniques but also because of its unique volcanic soil.

In the days of the Minoan civilization, Santorini was a dormant vol-cano, and vines were planted on antique lava and ash. Today, the vine-yard is entirely rooted in the layer of white pumice emplaced during the 1620 BC eruption: a layer tens of meters thick in most places, except on the steepest slopes, where the pumice fallout was unable to stick.

Pumice is a porous form of lava, ridden with small holes formed by the expansion of gas bubbles during the eruption process. Santorini's pumice is rich in silica, as well as in light metal oxides, such as aluminum, calcium, and sodium, hence the light color of the rock — which is classified as a rhyodacite.

Near the surface, Santorini pumice comes in a wide range of sizes, from popcorn to a particle of flour. When he showed me his estate near Oia, on the northern shore of the island, winemaker Paris Sigalas pointed out the advantage of this heterogeneous mix: it prevents the soil from setting like plaster and choking plant roots. Instead, the unsorted soil remains granu-lar and well aerated.

Another advantage of pumice is that it behaves like a sponge: the pores soak up the little moisture present in the air, including droplets of noctur-nal mist and windblown sea spray: winemakers credit the salty character of their white wine to this marine influence.

Santorini's pumice also has an interesting chemical makeup. It has low potassium content, which bolsters the wine's acidity: potassium usually acts to neutralize the acidity of grape juice, and its scarcity prevents this

Assyrtiko vine, growing on pumice from the 1620 BC eruption.

buffering effect. Hence, the wine's acidity, which translates into a crisp freshness, is directly related to the potassium-poor soil.

There are notable differences, however, from place to place, in terms of both soil and climate. When I asked Yiannis Roussos, owner of the Roussos estate, to comment on these differences, he answered sarcastically: "Is there not a difference between an elephant and a donkey?" And that was it.

Fortunately, his daughter Agape Roussos was more loquacious. Grapes from the central hills near Pyrgos, she points out, have a rather thin skin and yield a dry, fresh wine, whereas those closer to the seashore have a thicker skin, especially Aidani, and mature faster: they are harvested two to three weeks before those in the hills. Counterintuitively, despite their faster ripening, seashore grapes tend to develop subtler aromas.

Generalizing further, the northern half of the island has a more arid climate than the southern half, which translates into wines in the north (Sigalas estate) having more of a floral bouquet. In the south, where most vineyards are located, differences have more to do with altitude and proximity to the sea, as Agape Roussos pointed out.

Differences in altitude are often connected to differences in soil, as Stefanos Georgas showed me on his Argyros estate, on the outskirts of Episkopi village. Sheltered from the southern winds by a limestone ridge, his vineyard stretches over 35 hectares (86 acres) — the largest grouping of vine lots on the island — and shows interesting trends. On the upper slopes, one treads the thick white blanket of pumice of the 1620 BC eruption, whereas on the lower slopes, the pumice is studded with big chunks of dark basalt. This coarser, stony ground heats up slowly during the daytime and likewise takes more time to cool down after sunset. Wines from this terroir are described as strong and complex. By contrast, Georgas finds the grapes from the upper white slopes to yield more delicate, finer wines. But to truly experience the subtleties of the Santorini terroir, Assyrtiko should age several years in the cellar: it then becomes possible, according to the winemaker, to recognize the earthy notes typical of each particular place.

A Variety of Wines

Santorini boasts a gamut of original varieties and distinct vineyard settings. It also produces a range of different wines. Three types are distinguished by a Protected Denomination of Origin (PDO, which has replaced the original OPAP denomination) and are produced according to strict rules in order to ensure quality and authenticity: the sweet amber-colored Vinsanto, and the two dry whites Santorini and Santorini-Nykteri.

Santorini

By far the largest production on the island, Santorini dry white is predominantly Assyrtiko (75 percent minimum and up to 100 percent), eventually blended with up to 25 percent Athiri and/or Aidani. It has a bouquet of lemon, peach, and white flowers; a fresh bite provided by its acidity; and often a salty, mineral final note. To preserve this fresh, fruity character, Santorini is generally produced in vats of stainless steel, but some versions are at first briefly aged in oak. With time, oxidation can bring out notes of smoke, as well as lemon curd and walnut. Santorini dry whites pair well with Greek salads and feta cheese, shellfish, swordfish, and poultry and other white meat.

Nykteri

Santorini-Nykteri, or Nykteri for short, which means "night made," re-lies on the same Assyrtiko grape, with minor quantities of Athiri and/or Aidani. Harvested at full maturity, the sugar-rich grapes were tradition-ally fermented overnight — to take advantage of lower ambient tempera-tures — in closed vats, to limit contact with air and oxidation. Although the wine making no longer takes place at night, PDO regulations for Nyk-teri enforce a minimum of three months in oak barrels (though wine-makers often age it for six months), which add hints of vanilla, melon, and honey to the wine's citrus-fruit bouquet. On the palate, flavors of candied orange and lemon prevail.

Because it is particularly tannic, due to a longer maceration phase when it is in contact with the grape skins, Nykteri can be mistaken for a red wine in a "true" blind test (blindfolded, that is). Rich and complex, it pairs well with grilled fish, smoked meat, duck à l'orange, and strong cheeses. It ages well — up to fifteen years for the best vintages — and develops earthy notes.

Vinsanto

The third type of wine is Vinsanto, which uses the same three grape vari-eties: Assyrtiko (50 percent of the blend at minimum), Athiri, and Aidani. The late-harvested, mature grapes are spread out on bleached terraces and dried in the sun for eight to fifteen days. Dehydrated and saturated in sugar, the grapes are then pressed, and the maceration phase — when the juice is in contact with the grape skins — can last until Christmas. The re-sulting wine is then aged in oak barrels: it turns from golden to amber and even chocolate colored as it picks up more tannin from the wood. Before being bottled, the wine sits in oak for at least two years, and often longer: the label specifies the oak-aging duration in multiples of four — for example, four, eight, twelve, sixteen, or twenty years — with the price fol-lowing suit.

Vinsanto is served as an aperitif, as a dessert wine — it pairs well with strong cheeses and with chocolate and walnut desserts — or as an after-dinner drink, at a cool serving temperature of 6 to 8°C (43°F–46°F). It has a rather low alcohol content (10 percent by volume) and high residual sugar (200–300 grams per liter). Its bouquet typically blends orange zest and caramel, evolving over time toward chocolate and roasted coffee. In terms of flavor, exotic fruit also jumps in — pineapple, mango, and papaya — as well as honey, maple sugar, and crème brûlée.

Other Wines

Besides Santorini's three leading appellations, several other wines are marketed under the less restrictive label PGI (Protected Geographical Indication), namely the Aidani, Athiri, Mavrotragano, and Katsano varietals. Other island wines avoid classification altogether and reflect the winemaker's free spirit and desire to explore unbeaten paths.

A few blush wines are produced from red cultivars Mandilaria and Mavrotragano, often combined with Assyrtiko. Some are quite classy, like Argyros estate's Atlantis Rosé, which features a subtle bouquet of rose petals and melon. The best reds are Mavrotragano varietal wines, with a bouquet of dark berries, chocolate, and spices (Karamolegos, Sigalas, and Hatzidakis estates).

Red blends embrace different styles, such as Gavalas estate's Xenoloo, a mix of Mavrotragano, Voudomato, and white Assyrtiko that boasts a nutty aroma; or Santo Wines' Crescento, which blends Mavrotragano, Voudomato, and Mandilaria.

Rather than rely on descriptions, it is better of course to taste and drink these Santorini wines firsthand. They can be purchased in select wine shops, or on the Internet, either directly from producers or through wine importers. The other, better solution is to travel to Santorini: the island is one of the most beautiful spots on Earth, and the opportunity to visit should not be missed. Below is some advice to help in preparing a trip to Santorini, and also a list of wineries to visit.

Guide Section: Visiting Santorini

Santorini is a popular destination, especially during the summer, when prices are the steepest, and flights and hotels should be booked well in advance. Spring and fall are preferable, when there are fewer crowds and the weather is cooler. Even winter is an interesting season: islanders are less busy and easier to approach, and religious celebrations and feasts are a good occasion to get to know them. On February 2 (Orthodox feast day of the Ypapanti, or Candlemas), banquets are held in candlelit grottoes in the vineyards.

Santorini can be reached by plane, or by boat from Athens: the latter is an interesting option, allowing for a daylong cruise (eight hours) amid the Greek islands and a spectacular approach to the caldera from the sea. During the summer season, hydrofoils also operate to Santorini from Heraklion in Crete (a two-hour trip).

Santorini island is 18 kilometers long and 6 kilometers wide (11 by 4 mi.). Because of its size, it is well suited for scooters and bicycles (or even hiking), but a rental car is more practical. Most large hotels for groups are at Kamari on the eastern side of the island, close to the black-sand beaches. There are many small hotels and bed-and-breakfasts in the main towns of Fira and Oia, and along the caldera rim, closer to Akrotiri.

The many restaurants around the island serve excellent food, featuring fresh produce — Santorini goat cheese, tomato and eggplant, as well as seafood — best accompanied of course by a bottle of Santorini white wine.

Geology and Archeology Tour

Besides wine estates and the breathtaking postcard-perfect sites described in every guidebook, Santorini is a treasure trove for geology and archeology buffs.

Fira and Oia villages are worth visiting but teem with tourists. Below Fira lies a small harbor — reached by a mule path or by cable car — from which tour boats depart for the central Kameni islands. The boat ride offers a great view of the caldera walls from the sea, but the hike on Nea Kameni island is fairly uninteresting for volcano buffs: the view boils down to a couple of rubble-filled craters, with lazy fumaroles — sulfur-rich vapors —, that date back to the 1950 eruption.

Back on the main island, what is unique and truly worth seeing is the blanket of pumice from the 1620 BC eruption and the site that was buried under the fallout. One can tour these main archaeological sites in a day's time, starting in the morning with the Museum of Prehistoric Thira,[3] near Fira's bus terminal. The museum hosts many of the remarkable frescoes and other works of art recovered from the Minoan city of Akrotiri on the southern tip of the island.

On the way out of the town of Fira, south toward Akrotiri, there is a great exposure of pumice on the caldera rim. Park in front of the Carrefour supermarket and walk down the parking lot's exit to the gateway of an abandoned pumice quarry. A dirt road runs alongside a bank of pumice studded with blocks of basalt, violently ripped out of the volcano's chimney by the eruption. Down the path, on the north side of the bank (facing Fira), the sharp distinction is visible between the 1620 BC ash fall and the darker rubble below, which at the time was ground level.

The first 4 meters (12 ft.) of pinkish pumice were dropped by the ash column at the start of the eruption. Above that layer, the pumice turns finer and white, peppered with chunks of basalt: it was laid down by ground surges of hot, wet ash (the "tuff ring" phase) released when seawater entered the system and flashed to steam, propelling a base surge of debris in all directions, like the ground-hugging ring of an atomic bomb.

The later phases of the eruption are not recorded at this site — their deposits were carved off by erosion — but they are preserved elsewhere on the island. Drive south to Akrotiri (a fifteen-minute drive) and take the left fork in the road that leads down to the seaside archeological dig.

There is a parking lot near the entrance, and before visiting the digs, walk down to the coastline. To the right is a small wharf lined with taverns; to the left, past a seaside hotel, the ash cliff resumes the story of the eruption: the lava-studded pumice is topped by a thick layer of fine white ash, sculpted by erosion into a honeycomb of alcoves and ribs. The fine powder of this layer probably fell from the roaring ash column during the climax of the eruption and spread down the slopes as a hot pyroclastic flow.

On the wharf side of the beach, cellars were dug into the pumice cliffs when the archeological digs began in the 1960s. Several of these were later turned into taverns — Nikola's was the first and arguably is still the best — that serve Greek salads and seafood, and Santorini house wine.

From the small wharf, shuttle boats leave for the nearby lava beaches (Red Beach and White Beach). One can also reach them by a fifteen-minute hiking trail that starts from the parking lot opposite the digs, or directly by car via the modern town of Akrotiri.

The Akrotiri archeological site is shaded by a "bioclimatic" green roof.[4] A boardwalk circles the 1 hectare site (2.5 acres), with ramps leading down to selected sections of the Minoan city. One should note in particular the earthquake-split stone staircase in the Delta complex, near the center of the dig, and layers of pumice from the eruption, left standing along several walls.

Wine-Tasting Tours

There are a dozen independent wine estates on Santorini open to the public, as well as the large Santo Wines cooperative. Because of the island's small area, one can visit all the wine cellars in a couple of days. Several itineraries are listed below. They are grouped geographically, but any lineup is possible, as it takes less than half an hour to drive from any one wine cellar to another.

Kamari to Oia, the Cross-Island Tour: Gaia, Hatzidakis, Sigalas

Start the day on the beach, facing the morning sun, and end it watching the sunset on the west coast, with three wine estates in between. This cross-island itinerary begins at Kamari, circles around the hill of Pyrgos and heads north to the picturesque town of Oia on the caldera rim.

The first wine stop, before lunch, should be Gaia Wines, near the airport.[5] The estate took over an old brick-walled tomato cannery on the beach. Cofounded by Yiannis Paraskevopoulos—a rising star of Greek wine making—Gaia Wines produces three Assyrtiko dry whites, including one that highlights the terroir of the Pyrgos hills.

After this first round of wine tasting, head for Kamari village, which offers a vast choice of taverns and restaurants.[6] Following lunch, the second estate to visit—Hatzidakis winery—is ten minutes by car, in the hills above Kamari, near Pyrgos village.[7] Like most estates in Santorini, it is important to arrange for your visit ahead of time. Organic farming and rigorous wine-making methods deliver classic Vinsanto, Santorini-Assyrtiko, and Nykteri, as well as several "terroir" versions highlighting specific Pyrgos hillsides and old vines, such as Assyrtiko de Louros and Assyrtiko de Mylos. Hatzidakis also produces varietal wines: an Aidani white, a Mavrotragano red, and a Voudomato sweet red.

Leaving Santorini's central hill of Pyrgos, head north past Fira, across a limestone ridge, and down toward the coastline to Domaine Sigalas.[8] A former math teacher, fluent in several languages, Paris Sigalas is passionate about his island, vineyard, and history, and he runs a well-organized wine-tasting outpost in the middle of his estate, with professional and friendly staff serving a variety of Santorini-Assyrtiko and Nykteri dry whites, Aidani and Mandilaria varietals, and several blends. You can watch the sun go down over the vines from the wine-tasting terrace, or (with a designated driver) move on to nearby Oia, to watch the sun set over the sea.

Near Megalochori: Boutari, Gavalas, and Santo Wines

South of Fira, three estates are located near the caldera rim and represent three aspects of Santorini's wine-making history: the modern Boutari winery, the traditional Gavalas estate, and the Santo Wines cooperative.

Outside Megalochori village, a narrow road leads to the modern Boutari winery, crowned by a large white dome, near the caldera rim.[9] With its headquarters in Thessaloniki on the mainland, Boutari developed its

Santorini branch in 1989, hiring some of the island's best winemakers. Boutari wines range from classic blends to more experimental varietals. Besides Crete and Naouma whites and reds from outside the island, the estate produces Santorini-Assyrtiko and Nykteri whites, and a Mandilaria red varietal.

After this first stop, return to the main road, drive north to Megalochori and follow signs on the right to a narrow road leading to the Gavalas estate.[10] Its traditional *canava* winery dates back to the nineteenth century, with a family-like atmosphere and an instructive tour of the cellar, including an ancient grape-stomping stone floor. The wine tasting, in the shade of the courtyard, proposes a full range of Vinsanto, Santorini-Assyrtiko and Nykteri whites, as well as the varietals Aidani and Katsano (whites), Voudomato (rosé), and Mavrotragano (red).

To end the afternoon, return to the main road and head north out of Megalochori toward Fira. At the intersection with the road that leads inland to Pyrgos, the entrance to Santo Wines is on the cliff side, with a large parking lot and an aesthetically designed building complex, flush with the landscape, that includes an underground winery, a central shop and a wine bar, and outdoor terraces overlooking the caldera.[11] Besides Vinsanto and all the classic whites, red blends, and varietals, Santo Wines also processes tomatoes, capers, and fava beans, and offers a wide range of products in its well-stocked shop. One can tour the premises, taste a variety of wines (between four and eighteen types), and take in the breathtaking view, including the sunset, from the outdoor terraces.

Around Meso Gonia: Wine Museum, Karamolegos, Argyros, and Roussos

Between the airport to the east and Pyrgos hill to the west, Santorini's central plains host several interesting wineries, starting with Volcan Wines in Vothonas, where Georgios Koutsoyannopoulos set up a wax museum in his large underground cellar dedicated to the history of wine making on the island.[12] Ancient tools on display include a winepress brought by monks from southern France in 1660. The visit ends with a wine-tasting session of at least four different wines: whites, reds, and Vinsanto.

A short distance from the museum, in the village of Exo Gonia, the estate of Artemis Karamolegos, one of the rising stars of Santorini wine making, is also well worth a visit, best scheduled around lunchtime. In addition to the winery visit and wine tasting, a restaurant on the premises overlooks the vineyard and uses fresh produce from the estate's garden, paired of course with the wines: Santorini-Assyrtiko, Nykteri, Vinsanto, and varietals Aidani, Mandilaria, and Mavrotragano.[13]

The nearby village of Episkopi Gonia (also called Meso Gonia) hosts two of the island's oldest estates, located a couple of blocks from each other on the main street: Argyros and Canava Roussos. Estate Argyros has an intimate setting, is famous for its Vinsanto, and also produces excellent Santorini-Assyrtiko and Nykteri, varietals Aidani and Mavrotragano, and white, red, and rosé blends.[14]

Down the street, Canava Roussos offers a tour of its winery.[15] Its shaded wine-tasting terrace is ideal for large groups, with wines by the glass covering all classic Santorini productions, as well as several varietals and blends. The estate organizes dinners and concerts: the perfect way to end the Santorini experience.

CHAPTER 3

Mount Vesuvius

Mount Vesuvius is known for its explosive eruptions, most famously the burial of Herculaneum and Pompeii under a shroud of ash in AD 79. The volcano is also one of the most prolific fruit-growing areas in Europe, known for its apricots, tomatoes, and vineyards. Wines from Mount Vesuvius are marketed the world over under the appellation Lacryma Christi: the "tears of Christ."

Towering over the Bay of Naples, Mount Vesuvius has impressed Western consciousness more than any other volcano, since it was responsible for the cataclysmic eruption that destroyed Pompeii, Herculaneum, and Stabiae in AD 79. It is not so much the number of casualties that is striking—1,100 bodies unearthed so far in Pompeii—but the fact that entire cities were buried and uncovered fifteen centuries later, revealing past Roman life in dramatic detail.

Besides the tragic eruption of AD 79, Pompeii and its region have a wider story to tell. They show how a volcano can bring prosperity to a land by providing fertility and a microclimate: reason enough to attract a population to settle on its slopes, despite the damage caused by its moody behavior.

The eruption of 79 was not the first of its kind, nor will it be the last. Another cataclysmic eruption occurred in 1631, killing an estimated four thousand people, followed by a string of minor eruptions over the past three centuries and a final blowout in March 1944. Since then, the lull in activity is more disquieting than reassuring. The consensus among vol-

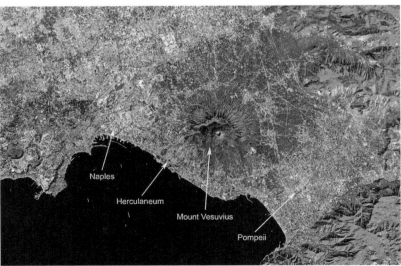

Location (top) and satellite image of Mount Vesuvius and the Bay of Naples, Italy.

canologists is that one eruptive cycle is over and another is brewing. The longer the respite, the more violent the reawakening will be: the magma chamber at depth is recharging with water vapor and carbon dioxide, volatiles that will ultimately set off ash clouds and pyroclastic flows.

The Bay of Naples

Nearly two dozen townships, totaling seven hundred thousand residents, live within 10 kilometers (6 mi.) of the summit crater and are within reach of deadly pyroclastic flows, be it modern-day Pompeii, at the edge of the archeological site; Ercolano, which grew back atop the buried ruins of Herculaneum; or the bustling agricultural towns of Boscoreale and Trecase. Naples itself, with a population of eight hundred thousand in the city proper, and close to three million including the suburbs, lies 20 kilometers (12 mi.) west of the summit, under the threat of ash fall if easterly winds blow during an eruption. In addition to many casualties, a major eruption of Mount Vesuvius today would destroy agriculture and ruin the economy of much of Campania, one of Italy's most fertile provinces.

Despite the threat, the population of the Bay of Naples has no intention of deserting its volcano. The lower flanks not only are fit for agriculture, because of the texture and mineralogy of lava flows and ash falls, but also offer a variety of microclimates based on elevation and orientation that fit the needs of different crops. The lower slopes long exported fruit and vegetables across Italy and even abroad, until the trade unification of Europe unleashed foreign competition and terminated the golden age of Mount Vesuvius. Today, fruit and vegetable farms cater mostly to local markets, with three exceptions still famous nationwide — tomatoes, cherries, and apricots.

One offering has never faltered: Vesuvian wine. The region of Campania is a top-rated wine-making province, with a reputation that dates back to the Roman Empire. Within Campania, Mount Vesuvius occupies a special niche because of its outstanding soil and microclimate, which earned it a wine appellation of its own: Vesuvio DOC.[1] And within that appellation, the top-rated wines — originating from twelve townships, reaching 12 percent of alcohol by volume (ABV), and relying on a select list of local grape varieties — earn the coveted label of Lacryma Christi ("tears of Christ"), wines that are served in Italian restaurants the world over.

These are the main reasons Mount Vesuvius is a top destination, not only for its appeal in terms of volcanology and archeology but also for the important role that it plays in gastronomy and wine making. If there is one place where all these stories come together, it is the ancient city of Pompeii itself, where new vineyards are rising from the ashes.

The Vineyards of Pompeii

It was a city of nearly twenty thousand souls — about half the population of Atlantic City or Palm Springs today — and covered close to a square kilometer in area (nearly half a square mile). Founded in the seventh century BC by the Oscan tribes of central Italy, built up by the Greeks, then by the Samnites, Pompeii fell under Roman rule in 80 BC and became a major trading post, as well as a resort for Roman nobility. It owed its reputation as much to its pleasure-oriented lifestyle as to its booming economy, in which wine making played a leading role.

The Pompeiian way of life is known to us in great detail, as a result of the city's burial under 7 meters (22 ft.) of ash: a tomb that was impossible to plunder. Pompeii's very location was forgotten until the architect Domenico Fontana chanced upon the buried walls in 1599 while digging a water inlet for a factory nearby. It would be decades before archeologists realized that the ruins were those of Pompeii, and official digs did not start in earnest until 1748. Block by block, the city was stripped of its ash cover. Nearly three centuries later, two-thirds of the city has been unearthed — the last section is still hidden under its seal of pumice.

Pompeii was ideally located on the edge of a lava plateau, 30 meters (100 ft.) above sea level, with a good view over the Bay of Naples, and the coming and going of merchant ships. The Sarno River flowed past the city walls, connecting Pompeii upstream to the agricultural hubs of Nola and Nocera, and downstream to the Mediterranean and its rich fishing grounds.

In addition to controlling trade routes, Pompeii possessed its own prized cropland. Pumice and broken-up lava flows formed a lightly packed, mineral-rich soil at the foot of gentle slopes facing the rising sun and sheltered by the volcano from the rain-carrying northern winds. Pompeii's rich merchant class owned vineyards and olive trees on the flanks of Mount Vesuvius, and inside the city as well, covering nearly a third of its area.

The reputation of Pompeiian and Campanian produce was widespread throughout the Roman Empire. In his *Natural History* encyclopedia, Pliny the Elder, who met his tragic death in the AD 79 eruption, described the land as follows: "From this gulf onwards stretch forth vine-covered hills, well known throughout the world for the elation that their illustrious nectar provides, not to mention, as our Ancestors would say, the supreme rivalry between Liber Pater and Ceres—between wine and wheat."[2]

Wine making was already an advanced art form in the first century AD, and in his chapter on wine, Pliny lists no fewer than ninety-one grape varieties known to the Romans, as well as pruning and farming practices, wine-making techniques, and different types of wine.

In Pompeii, wine could be found everywhere, as one can sense by touring the ruins. Entering through the southeastern gate of the site—the Porta Nocera—visitors are directly confronted with the city's wine-making past, and for good reason. Here lies the Amphitheater district: the great open-air stadium that hosted circus games, with a capacity of twenty thousand people—the entire city population. Around the stadium, vineyards sold wine directly to spectators.

One could also drink wine at the theater—there were two theaters in Pompeii—as well as in the public baths while taking a dip, and of course in street taverns. Nearly two hundred taverns have been unearthed so far. They are known as *thermopolia*, with sidewalk-facing counters that served wine and hot meals.

In one tavern in particular in Herculaneum, writing on the wall lists the going price for wine: one as (a small copper coin, worth a quarter of a brass sesterce) for a cup of house wine; two asses for higher-quality Pompeiian wine; and four asses (one sesterce) for an imported "Grand Cru," most likely from Salerno—a famous wine-making district to the southeast.[3]

The Romans, as a people, drank a lot. Estimates place their yearly consumption of wine at 130 liters per person: much more than what the Italians or French drink today (50 liters per year).

Wealthy estates outside Pompeii had their own vineyards, as did the

Villa of the Mysteries (Villa dei Misteri), a quarter mile northwest of town. Besides its winepress, the villa is famous for its beautiful frescoes that depict the initiation of a young woman into the cult of Dionysus, god of the grape harvest and of wine, fertility, and ecstasy.

Known as Dionysus, Bacchus, or Liber Pater (in Greek, Roman, and Italic cultures, respectively), this important wine god is omnipresent in Pompeii, in the form of marble statues, bronze statuettes, and frescoes. One fresco in particular shows the god dressed in a cluster of grapes, standing in front of Mount Vesuvius, with pergola-trained vines covering the lower slopes.

It has often been written that the Pompeiians were unaware that Mount Vesuvius was a volcano until it came to life by surprise in AD 79. This might have been true for most of the population — there had been no sign of activity on the mountain for over three centuries — but the more literate citizens would have known better if they read their classics. Indeed, the Greek geographer Strabo (63–14 BC) pointed out in his writings that the summit of Vesuvius "looks ash-coloured, and shows pore-like cavities in masses of rock . . . as though they had been eaten out by fire; and hence one might infer that in earlier times this district was on fire and had craters of fire."

Strabo went on to speculate that "perhaps, too, this is the cause of the fruitfulness of the country all round the mountain; just as at Catania [Sicily], it is said, that part of the country which had been covered with ash-dust . . . by the fire of Etna made the land suited to the vine; for it contains the substance that fattens both the soil . . . and produces the fruits."[4]

Many Pompeiian winemakers, familiar with Mount Etna, had certainly jumped to the same conclusion, attributing the fertility of their terroir to the ash and lava, and recognizing Vesuvius as a volcano. Another giveaway was the building material used in Pompeiian villas, street pavements, and sidewalks. In his treatise on architecture, Vitruvius (90–30 BC) links the light pumice, abundantly used in the walls of houses, to the fires of Vesuvius, and the dark-green paving stone comes from an old lava flow underlying Pompeii.

Pompeii actually commercialized its lava. Known to geologists as trachyte, it is studded with crystals of sanidine — a tabular, vitreous gray mineral — and is both elegant and highly resistant: an ideal material for grinding stones used to press olives for oil or to crush wheat for flour. Such presses featured two nested lava cones that crushed wheat or olives along their interface once the inner cone was set in motion.

Initially, Pompeii was very successful in marketing lava grinding stones throughout the region. In his treatise on agriculture, Cato the Elder (224–139 BC) praises their quality and even mentions their price: 724 sesterces for a top-of-the-line model, including shipping and assembly.

During the decades leading up to the eruption in 79, however, there is evidence that the Pompeiian grinding-stone industry was losing ground. In the thirty-four bakeries uncovered in Pompeii so far, grinding stones made from Vesuvian trachyte were set aside, out of commission. Those in working order were made from different lava: phonolite from Orvietto, a trading town north of Rome, with quarries conveniently located for shipping on the Tiber River. Were they cheaper or more fashionable? Either way, Pompeii was losing the grinding-stone market to Rome.

As it were, Pompeii was not faring well in its final years. A powerful earthquake rocked the city in 62, some seventeen years before the fatal eruption. Many buildings were badly damaged, including the market next to the forum, the main theater, and most public baths, as well as a water tower and aqueducts servicing the city.

No one knows whether the citizens of Pompeii, busy as they were repairing their homes, made any connection between the earthquake and a possible reawakening of Mount Vesuvius. As viewed from the city, the mountain appeared to be at a safe distance, roughly 8 kilometers (5 mi.) away. But when Vesuvius roared back to life, the city proved much too close.

The Fatal Eruption of AD 79

On the morning of August 24, AD 79, a cloud of steam rose above Mount Vesuvius as fresh magma reached the surface, vaporizing groundwater along the way. Downwind of the volcano, a sprinkle of fine ash began to fall from the sky.

By noontime, the eruption gathered momentum. We have a good understanding of the sequence of events from combing through the stacked layers of ash and scoria that settled over Pompeii, and also thanks to Pliny the Younger (AD 61–114), nephew of Pliny the Elder, who was eighteen at the time and viewed the eruption from across the Bay of Naples. As he later wrote to historian Cornelius Tacitus:

> On the 24th of August, at about one in the afternoon, my mother [called my uncle] to observe a cloud which appeared of a very unusual size and shape. . . . A cloud, from which mountain was uncertain at this distance (but it was found afterwards to come from Mount Vesuvius), was ascending, the appearance of which I cannot give you a more exact description than by likening it to a pine tree, for it shot up to a great height in the form of a very tall trunk, which spread itself out at the top into branches . . . ; it appeared sometimes bright and sometimes dark and spotted, as it was more or less impregnated with earth and cinders.[5]

The eruption was a cataclysmic discharge of pumice and finer ash, caused by the violent release of gas stored in the magma chamber: a phenomenon known today as a Plinian eruption, in reference to the historic description by Pliny the Younger. From the aerial extent of the fallout, geologists estimate that the eruptive column rose to a height of at least 15 kilometers (roughly 10 mi.), spreading out horizontally when it hit the jet stream: hence the comparison to a pine tree.

The city of Pompeii lay directly downwind of the volcano. It was hit by a dense shower of pumice, some the size of chestnuts, that began to pile up in the streets and on the roofs and terraces. By three in the afternoon, two hours after the onset of the eruption, the blanket of pumice was already 30 centimeters (1 ft.) thick.

Pompeiians were faced with two alternatives: stay put, to protect their material possessions, or leave the city at once. Those who chose to flee and took to the road—or piled into the small boats moored on the Sarno River—were well inspired. Time was of the essence, as conditions were worsening by the minute. As the ash fall thickened and darkness spread over Pompeii, the fugitives were soon groping in the darkness, their torches choked by swirling ash.

We can picture the scene, based on Pliny the Younger's second letter to Tacitus, in which he describes his own plight as the eruption cloud engulfed Cape Misena, west of the volcano, forcing him to join the panicked crowd:

A dense, dark mist seemed to be following us, spreading itself over the country like a cloud. . . . Night came upon us, not such as we have when the sky is cloudy, or when there is no moon, but that of a room when it is shut up, and all the lights put out. You heard the shrieks of women, the screams of children, and the shouts of men; some calling for their children, others for their parents, others for their husbands . . . ; some wishing to die, from the very fear of dying; some lifting their hands to the gods; but the greater part convinced that there were no gods at all, and that the final endless night, of which we have heard, had come upon the world.[6]

We know from the number of corpses that many Pompeiians chose to stay or were trapped. By five in the afternoon, roofs began to cave in under the weight of the ash blanket, claiming the first victims. Two thousand years later, stripping the homes of their shroud of pumice, archeologists are still discovering bodies crushed under beams and collapsed rooftops.

By eight in the evening, the pumice had reached a thickness of 2 meters (6 ft.), blocking all ground-floor doors and windows. Its color had switched from white to gray, signaling a change in composition of the magma and a further jump in the intensity of the eruption. From the increased extent of the gray fallout, the billowing cloud must have reached an altitude of 30 kilometers (roughly 20 mi.).

The Final Blow

The worst was yet to come. The thickening ash column over Mount Vesuvius had reached such a high density that its buoyancy began to falter. At one in the morning on August 25 occurred the first collapse of the ash column under its own weight, sweeping down the slopes of the mountain in what is known today as a pyroclastic flow. The avalanche of scalding gas and pumice funneled down into the valleys and gullies, flanked by side wings of hot air and fine ash that were able to jump over topographic obstacles — a shock wave known to volcanologists and atomic-bomb specialists as a base surge.

The first pyroclastic flow headed south, toward the coastal town of Herculaneum. The glowing pumice slid around the city walls and out to sea, but the flanking base surge of hot gases blew over the obstacle of the city, incinerating the population.[7]

This first pyroclastic flow missed Pompeii—out of range to the east—but the ash column kept building up over the crater, repeatedly reaching the threshold of collapse. A second pyroclastic flow blew over Herculaneum, and a third came dangerously close to Pompeii around six in the morning, dropping a bed of hot ash along its western walls.

The fourth pyroclastic flow, an hour later, at seven, was the fatal one: its shock wave blew over the entire city and finished off the population. Of the 1,100 victims uncovered in the ruins so far, one-third are thought to have died during the afternoon, evening, and late night of August 24 from falling rocks and roof collapse, and two-thirds from the 7 a.m. pyroclastic flow.

Two more fiery ash flows, even larger in scope, blew over the city during the following hour, toppling walls and bringing the total thickness of ash and pumice to a full 7 meters (23 ft.).

The shock wave from the final surge rolled over land and water as far as Stabiae, 10 kilometers (6 mi.) south of Pompeii. Its most famous casualty was Pliny the Elder, the author of the famous encyclopedia *Natural History* and admiral of the Roman fleet. He had landed the previous evening near Stabiae in an attempt to rescue friends. After spending the night under shelter, surrounded and nearly trapped by the falling pumice, Pliny and his crew finally made a run for the beach to reembark at dawn, but the party was overtaken by the morning surge of hot gas: the elder statesman suffocated, but several members of his party survived to tell the story.

All in all, the climax of the eruption lasted twenty-four hours. Mount Vesuvius continued to spew ash and feed minor pyroclastic flows over the following few days, then scaled down its activity to within the central crater before shutting down completely. Throughout the next millennium, minor eruptions shook the mountain every century or so, with few, if any, casualties.

The trauma of the eruption in 79 was deep and lasting. Pompeii was never rebuilt and stayed buried under its shroud of gray pumice, until its very location sank into oblivion. Sealed off from the atmosphere, it became a time capsule of Roman life, waiting to be reopened.

POMPEII'S BURIAL DATE REVISITED

Experts debate the exact date, in AD 79, of the eruption of Mount Vesuvius and the destruction of Pompeii. According to the description by Pliny the Younger, it was "the ninth day before the calends of September," which translates in our calendar to August 24. However, Pliny the Younger wrote his account twenty-seven years after the fact (in AD 106), and his memory might have failed him, or else the date was mistranscribed when the original letter was later copied.

Several archeologists involved in the digs believe instead that the eruption took place a couple of months later than Pliny's account, perhaps on the ninth day before the calends of November, on or about October 24. They base this belief on the fact that castings of victims entombed in the ash show them wearing thick clothing rather than light summer tunics. Another clue is a coin featuring the effigy of the emperor Titus, depicting his fifteenth imperial salutation—a celebration that allegedly took place later than August. But the most convincing evidence for a fall eruption rather than a summer one comes from the fruits and nuts discovered in the ruins.

Among the fossilized produce identified in the pantries and stores of the city, fruits typical of the month of August were in dried form. Those that were fresh were typical of October: olives, figs, and chestnuts. Last, Pompeiian wine brings its own lighting to the debate: amphorae of the 79 vintage were already sealed, an operation that usually took place in October. *In vino veritas?*

Castings of victims suffocated by the AD 79 eruption, at the ruins of Pompeii.

A Vineyard Rises from the Ashes

The pumice and fine ash that buried Pompeii preserved many frescoes, statues, furniture, and daily objects. They also "froze" for posterity crops and vineyards across the city: not only seeds and other hard parts that resisted heat and alteration, but also softer plant tissue that was pyrolyzed and dissolved, leaving hollow spaces in the hardened ash. Archeologist Giuseppe Fiorelli, in charge of the dig in the 1860s, developed an ingenious technique to recover the missing information: he injected liquid plaster into the cavities, creating molds of animal and human victims, as well as of tree trunks and vine stock.

In the many vineyards buried in ash, the injected plaster materialized the volumes once occupied by trunks and roots — one way to figure out which grape varieties were planted at the time. As a result, archeologists and winemakers came up with the idea to replant vines in Pompeii, in exactly the same locations they occupied two thousand years prior, and using identical grape varieties. The experience was interesting from a historical perspective, but reestablishing vineyards in Pompeii would also cheer up the ruins and bring life back to the city.

In charge of the project, Pompeii's Applied Research Laboratory did not have to look far to find the perfect manager for the task. Based in Atripalda, an hour's drive from Pompeii, the Mastroberardino wine company was well known for growing and promoting ancient grape varieties.

Figuring out which grape varieties were in use two thousand years ago in the Roman city called for clever detective work: looking at plaster moldings of root systems, analyzing fossilized grape seeds recovered from amphorae where they were macerating, scrutinizing representations of grape clusters on Pompeiian frescoes, and reading the descriptions of popular varieties in the encyclopedia of Pliny the Elder.

From the combined evidence, the grape detectives managed to identify the red grapes Aglianico, Piedirosso, and Sciascinoso, and the white grapes Caprettone, Coda di Volpe, Falanghina, Fiano, and Greco, most of which are still used in the making of Lacryma Christi wine today.

Starting in 1996, the eight varieties were planted experimentally in Pompeii on the site of an ancient vineyard, the Foro Boario, next to the grand stadium (the Amphitheater).[8] The Mastroberardino team replicated Roman vine-training techniques, using chestnut stakes to train the red grapes and a pergola arrangement to train the whites. The soil — predominantly made up of pumice and lava fragments from prehistoric eruptions — was tilled once, to revive its texture and functionality (after two thousand years of idleness).

In 1999, three years after the initial planting, grapes were harvested for the first time and treated to a round of experimental wine making. Of the eight varieties tested, only two did well on the Pompeiian terroir and lived up to present-day criteria in terms of wine quality: Piedirosso and Sciascinoso.

Hence, the winemakers focused on the two varieties, planting them in other Pompeiian orchards that were known to have hosted vineyards in Roman days. They bottled their first official wine in 2001, after aging it in oak barrels (rather than terra-cotta amphorae, for quality's sake), and released it two years later in a public auction: all 1,721 bottles were sold off at an average price of €100 a bottle — a revenue partly reinvested in the archeological digs and restoration work pertaining to the vineyards.

Composed of 90 percent Piedirosso and 10 percent Sciascinoso, the first vintage has a deep ruby color, a fruity and spicy bouquet, a thick but balanced structure, and an alcohol content of 13.5 percent (ABV). Technically, it falls in the regional category Pompeiano IGT,[9] a designation that includes all wines from Mount Vesuvius and the Sorrento Peninsula.

Mastroberardino named its creation Villa dei Misteri, in reference to the Pompeiian estate and its frescoes that depict the initiation of a young woman to wine drinking. Fittingly, one of the frescoes is featured on its label. Today, the blend has switched from predominantly Piedirosso to a mix of 40 percent Piedirosso, 20 percent Sciascinoso, and 40 percent Aglianico, and a bottle of Villa dei Misteri continues to fetch around €100 on the web or in Naples airport duty-free shops.

Vesuvius Grapes and Terroir

On one of my visits to Pompeii, Mastroberardino's agronomist in charge, Antonio Dente, took me to the original experimental lot behind the Amphitheater, then a couple of blocks over to the orchard of a rich Pompeiian villa: Oste Eusinio. Here, the wine company had set up a grape conservatory for its eight antique varieties, including the two red leaders of its Villa dei Misteri wine: Piedirosso and Sciascinoso.

Piedirosso has a russet-colored stem, fanning out in three branches that resemble a pigeon's foot, hence its name Piedirosso ("red foot") or Palombina ("little dove"). In the days of Pliny the Elder, it was known as *Columbina purpurea* in Latin ("purple dove"). The naturalist mentioned that it bore particularly large grape clusters. As for Sciascinoso, Pliny and his contemporaries knew it as *Vitis oleagina*, because the small oblong grapes look like olives — hence its nickname, Olivella, in the Naples area today.

Another row of the conservatory features the white grape Coda di Volpe (*Caude vulpium* in Latin, or "fox tail"), which bears a long and bushy, tail-like grape cluster. It too has remained popular to this day and is one of the main ingredients in white Lacryma Christi wine.

Then there is Greco: as one might guess, it was introduced to the area by the Greeks. The Roman agronomist Lucius Columella (AD 4–70) described it as "very well known, because it covers those most famous slopes of Mount Vesuvius."[10] Greco is used in white Lacryma Christi today, as are Falanghina and Caprettone, also planted in the conservatory.

Last but not least, Aglianico (or *Vitis hellenica* in Latin), also brought to Naples by the Greeks, is a component of both red and rosé Lacryma Christi today, and Dente raised it to become the third component of his Villa dei Misteri wine, alongside Piedirosso and Sciascinoso. It took some effort: the first plantings did not do too well on Pompeiian pumice, with the grapes swelling and bursting before reaching full ripeness. Dente then grafted cuttings of Aglianico onto the sturdier Piedirosso rootstock, which proved to be the solution, as the latter is well adapted to the volcanic terroir.

Pompeii and its area are indeed blessed with a very special terroir — a notion that was already recognized in Roman times. In his *Natural History*, Pliny the Elder emphasized that "what matters is the land, and not the grape itself" and that a wine "is worth as much as the soil that it grows upon."[11]

The Romans also knew to vinify each variety separately rather than blend grapes outright. Lucius Columella recommends such varietal wine, in that there "is a great advantage to be able to preserve the flavour of every variety — not blended but true and genuine — and to put it away by itself. . . . Varieties, when racked off in this fashion, attain the rank of nobility, because nothing of a different sort enters in to counteract their keeping qualities." The agronomist also underscores the importance of proper ori-

MOUNT VESUVIUS: TECTONIC SETTING
AND MAGMA COMPOSITION

Mount Vesuvius is the southernmost volcano in a lineup known as the Roman Comagmatic Province. Farther south, the Aeolian Islands off the coast of Sicily and Mount Etna belong to a separate group.

The tectonic framework of these volcanic provinces is rather complex, because the Mediterranean basin is a patchwork of microplates, crushed between the large, converging African and Eurasian plates. Slabs of Mediterranean oceanic crust bend and sink back into the hot mantle, one piece under Greece (hence the volcanism of Santorini and the Hellenic arc), and another, including the Adriatic seafloor, under the Italian Peninsula. The latter plate dives obliquely under Italy, setting off plumes of magma downrange that rise under the peninsula's western, Tyrrhenian coast. The ascending plumes fuel, from north to south, the volcanoes of Sabatini, the Alban Hills, Roccamonfina, Ischia, the Phlegrean Fields, and Mount Vesuvius.

Mount Vesuvius was born approximately three hundred thousand years ago, according to the oldest lava flows at the bottom of the pile, which have been reached by drilling rigs. A first cone was completed roughly twenty-five thousand years ago. Initially higher than the present one, its outline remains visible as a horseshoe-shaped outer wall that encircles its successor, to the north and northeast. Still peaking 1,132 m (3,736 ft.) above sea level, this paleo-Vesuvius, known as Monte Somma, was cut down by a cataclysmic blowout seventeen thousand years ago, much more violent than the eruption in 79 that destroyed Herculaneum and Pompeii. In the center of the collapsed Monte Somma grew the modern cone (1,280 m, or 4,224 ft.), which has undergone several cycles of partial destruction and reconstruction, most recently in 79, 1631, and 1906.

The activity at Mount Vesuvius appears to be cyclical. Each cycle begins with a powerful Plinian eruption, followed by several centuries of lesser eruptions, and ends with a temporary dormant phase.

The chemistry of lava erupted at Mount Vesuvius is somewhat unusual. The Adriatic plate that dives under Italy not only melts some of the mantle but also undergoes melting itself, providing a crustal component to the magma. It is this crustal component—the meltdown of old lava flows and marine sediment—that enriches Vesuvian magma in potassium and makes the lava especially fertile.

entation of the vines: "those with a southern exposure yield a superior quality. And there is no doubt that the nature of some vines is such that in the quality of their wine they sometimes excel, sometimes are excelled, according to their situation."[12] In hot climes, in particular, Columella recommends an easterly exposure as more advantageous than a southern one.

Pompeii's location is a case in point: the vineyards lie on the volcano's southeastern slope, facing the rising sun. As for the substrate, it is porous scoria or "ash," laden with metal oxides. Chemical analyses show an enrichment of potassium, a fertilizing element traced to the volcano's tectonic setting. Potassium oxide in the substrate can reach up to 10 percent by weight, ranking third behind silicon and aluminum, which puts it on par with iron oxides.

For potassium and other metals to be accessible to plant life, they need to be removed from the crystal lattice of their host minerals. Chemical alteration by acidic runoff water transforms potassium-rich minerals (such as K-feldspar) into softer clay, releasing potassium and other ions into the water and making them readily available to plant roots.

The broken-down lava is first colonized by lichen, namely the coral-looking *Stereocaulon vesuvianum*; then the soil's metal ions and organic molecules are recycled by higher-order plants, including vegetables, fruits, and grapevines.

Italy's Orchard

Today, vineyards fan across the lower slopes of Mount Vesuvius in a south-facing crescent, from San Sebastiano to the west above Naples and out to Boscotrecase and Boscoreale to the southeast, above Pompeii. But there is more to Vesuvius than just wine: besides vineyards, there are olive groves, vegetable patches, and orchards. Citrus and peach trees are widespread, but the two most outstanding fruits are cherries on the northern slopes and apricots both north and south.

Vesuvian cherries are grown around the northern villages of Massa di Somma, Somma Vesuviana, Ottaviano, and Santa Anastasia. Two varieties stand out; they are sold fresh and also used in jams and ice cream. Malizia cherries are exceptionally large and a bit pointed, ruby red in color, with a tender, sweet pulp and spiked with a tangy aftertaste. Ciliega del Monte cherries are large as well, sweet and tangy, and double-faced: red on one side, and yellow or pink on the other.

Apricots number several dozen cultivars on Mount Vesuvius, the most famous being the sweet Pellechiella. Vesuvian orchards deliver nearly half of Italy's apricot production for fresh fruit, dried and canned fruit, jam and yogurt, as well as apricot brandy. Molecular analyses are under way to characterize and protect Vesuvian apricots under the Protected Geographical Indication (PGI) label. Their superior characteristics include high sugar content and pulp yield, uniform color, good texture, and rich flavor.

There is also a vine fruit, besides grapes, that achieves excellence on Mount Vesuvius: tomatoes. The variety is known as Pomodorino del Piennolo—*piennolo* means "hanging" in the Neapolitan dialect—because, once harvested, the tomatoes are hung in clusters to ripen under pergolas or in cellars. Oval shaped and with a pointed tip, they have a dense pulp, high sugar content, and a tanginess credited to the terroir's mineral salts. They also display a fiery red color, which they owe, according to legend, to the underground magma, directly tapped by the vine roots. Vesuvian tomatoes are prominent in Neapolitan cuisine—in fresh, dried, or crushed form—such as in pizza toppings, pasta, and sauces for meat and fish.

Despite such quality produce, Vesuvian agriculture is declining in the face of European competition: between 1960 and 1990, the area of cropland has been cut in half. Today's cropland covers only 40,000 hectares (100,000 acres), 60 percent of which are devoted to fruit production (not including citrus), 15 percent to tomato and vegetable crops, and 15 percent to vineyards. The balance (10 percent) includes citrus fruit, olives, and some cereal crops in the plains.

Large estates are disappearing, broken up into family-run smaller units or developed into urban zones. Villagers also grow a variety of produce in their backyard, making the best of their space by training grapevines on pergolas and freeing up the dark volcanic soil underneath to grow tomatoes, zucchini, broccoli, and strawberries in the shade: a clever two-tiered arrangement.

Villages and their cropland pepper the slopes of Mount Vesuvius up to 200 meters (660 ft.) above sea level. Individual homes, with their private orchards, extend a little farther up, until the gradient steepens noticeably. Here begins the prime wine belt that wraps around the volcano, between 250 and 350 meters (825–1,155 ft.) elevation, within reach of most lava flows that descend from the summit crater or from flank fissures lower down. Despite the threat from the volcano, this is where the best wine of the region is produced: Lacryma Christi.

Lacryma Christi: Tears of Christ

The Bay of Naples harbors many legends. The poet Virgil (70–19 BC) placed the entrance to hell on the shores of Lake Avernus, a water-filled crater west of Naples, in a volcanic area known as the Phlegrean Fields. In Roman times, birds flying over the crater were often seen to drop dead into its dark waters: perhaps the lake occasionally belched carbon dioxide from its depths, suffocating animals or even humans on its shores, as happened recently around similar crater lakes in Africa.[13] Today, the marshy lakeshores have been drained and planted with vines (Piedirosso, Aglianico and Falanghina), yielding their own appellation of Campi Phlegrei wines.

With the rise of Christianity, the area lost some of its hellish attributes, but the devil himself jumped in — literally. Expanding on the biblical account of Lucifer's fall from grace, Neapolitan folklore asserts that the falling archangel grabbed and ripped off a piece of heaven, carrying it to the ground, where it became the Bay of Naples. Witnessing the loss from above, Jesus Christ wept, and wherever his tears landed, wonderful vines sprang forth.

There are side stories to this legend. According to one, the tears of Christ are the sugar-rich droplets that ooze through the pores of overripe grapes. According to another, the tears are drops of alcohol that condense out of vapor during the fermentation process: a happy sign that grants the winemaker a place in heaven.

In any case, Lacryma Christi has enjoyed a reputation of excellence over many centuries. The French philosopher Voltaire, a fine connoisseur, mentions the wine in his novel *Candide* (1759); and in *The Count of Monte Cristo* (1844), Alexandre Dumas lists it as the favorite wine of one of his characters.

Technically speaking, Lacryma Christi is a subset of the broader Vesuvio DOC appellation that encircles the volcano. The authorized grape varieties are all indigenous to the area: Coda di Volpe, Caprettone, Falanghina, and Fiano for the whites; Piedirosso, Aglianico, and Sciascinoso for the reds and rosés. The yield must be restrained and the wines must top 12 percent alcohol to earn the Lacryma Christi label.

One of the better places to discover Lacryma Christi is Cantina del Vesuvio, an estate located off the main road to the summit crater, above the town of Boscotrecase. Owner Maurizio Russo realized that the idyllic setting of his estate — facing the Mediterranean and the island of Capri in the distance — would attract enough visitors to drive the family business, thereby introducing the novel concept of wine tourism to the area.

Plate 1. Santorini (Greece): The town of Oia, overlooking the caldera.

Plate 2. Santorini: The island of Nea Kameni and its recent lava flows.

Plate 3. Santorini: Minoan wine jugs, at the ruins of Akrotiri, 1620 BC.

Plate 4. Santorini: Staircase split by a volcanic tremor, Akrotiri, 1620 BC.

Plate 5. Santorini: Orthodox church and vines, on pumice from 1620 BC.

Plate 6. Santorini: Assyrtiko, the leading grape variety on the island.

Plate 7. Santorini wine types: Vinsanto, Nykteri, and Assyrtiko.

Plate 8. Santorini: *Kouloura* basket pruning, shown by winemaker Paris Sigalas.

Plate 9. Mount Vesuvius (Italy), viewed from the Cantina del Vesuvio vineyard.

Plate 10. Pompeii: Vineyard planted in the ruins of a Roman villa.

Plate 11. Herculaneum: Street tavern and wine bar (*thermopolium*).

Plate 12. Pompeii: Fresco of Mount Vesuvius, with wine god Bacchus at left.
National Archaeological Museum of Naples.

Plate 13. Mount Vesuvius: Lacryma Christi white grape (Caprettone).

Plate 14. Mount Vesuvius: Lacryma Christi red grape (Piedirosso).

Plate 15. Mount Vesuvius: Local tomatoes, pomodorini del piennolo del Vesuvio.

Plate 16. Mount Vesuvius: Tomatoes grown in volcanic ash, near the 1944 lava flow.

Plate 17. Mount Vesuvius: Winemaker Giordano Ciro (Cantine Olivella).

Plate 18. Mount Vesuvius: Winemaker Maurizio Russo (Cantina del Vesuvio).

Plate 19. Mount Etna, Sicily, towering over the city of Catania.

Plate 20. Mount Etna: Summit eruption and lava flow (2006).

Plate 21. Mount Etna: The 1991–1993 lava flow grinded to a halt in the suburbs of Zafferana Etnea.

Plate 22. Mount Etna: Virgin Mary and child, protecting the town of Zafferana Etnea.

Plate 23. Circumetnea train over vines and olive trees, with the volcano in the background.

Plate 24. Mount Etna: Nerello Mascalese, the main grape of Etna Rosso wine.

Plate 25. Mount Etna: The 1981 lava flow creates a special climate (Romeo del Castello estate).

Plate 26. Mount Etna: Grape harvest at Passopisciaro, on the northeastern slope.

Plate 27. Mount Etna: Winemaker Frank Cornelissen, in one of his high altitude lots.

Plate 28. Mount Etna: Winemaker Salvo Foti, in front of a century-old Nerello Mascalese.

Plate 29. Mount Etna: Citrus fruits are the number-one crop on the volcano's southern slope.

Plate 30. Mount Etna: Grown on the northern flank, Bronte pistachios are world famous.

Plate 31. Stromboli volcano blowing a puff of ash over a newly planted vineyard.

Plate 32. Sicily and Aeolian Islands (Y-shaped archipelago) to the north, from space.

Plate 33. Vulcano: La Fossa volcano, overlooking the harbor.

Plate 50. Canary Islands from space. Lanzarote is the island at the upper right.

Plate 51. Lanzarote: Timanfaya cinder cones, from the 1730–1736 eruption.

Plate 52. Lanzarote: Wine tour of Bodega La Geria, among vines planted in ash.

Plate 53. Lanzarote: Bowls dug in ash (*hoyos*), protected from the wind by stone walls.

Plate 54. Napa Valley: Volcanic Vaca Mountains overlooking Fay and Stag's Leap Vineyards.

Plate 55. Napa Valley: Stag's Leap basalt ledge, reflecting afternoon light on Fay vineyard.

Plate 56. Cabernet Sauvignon, the ruling grape of Napa Valley.

Plate 57. At the source of Oregon's terroir are lava flows, scarred by Ice Age floods.Plate 57.
At the source of Oregon's terroir are lava flows, scarred by Ice Age floods.

Plate 58. Willamette Valley: Cristom Estate, in Eolia–Amity Hills, on Nekia volcanic soil.

Plate 59. Hawaii Big Island: Eruption at Leilani Estates, March 2018.

Plate 60. Hawaii Big Island: Coffee plantations, Kona coffee belt.

Plate 61. Hawaii Big Island: Kona coffee berries.

Whereas Maurizio's grandfather used to load his mule-drawn carriage in the middle of the night to reach downtown Naples in time for the morning market, it was now the customer who would make the trip up the mountain to buy the wine.

Maurizio and his wife, Rossana, welcome visitors on the open terrace flanking their farmhouse, with tables set in the shade of olive trees and freshly baked bread, garden tomatoes, cheese, and olive oil served along with the wine. The day of my visit, Maurizio had to serve lunch to a busload of tourists, answer my questions and his cell phone, and watch over an Italian food critic tasting and grading his wines for her yearly guidebook.

After a dry *spumante* for starters, Cantina del Vesuvio offers white, rosé, and red Lacryma Christi. The white is a blend of 80 percent Coda di Volpe and 20 percent Falanghina. Straw colored, and giving off a bouquet of flowers — reminiscent of the blooming brooms on the volcano — it is a dry, slightly tangy wine with a peachy aroma, a touch of quince, and a hint of pineapple. Lacryma Christi white is best served with shellfish, a seafood risotto, or, if one happens to be in Naples, an *impepata di vongole*: clams cooked in Vesuvian tomatoes and black pepper.

The Cantina's Lacryma Christi rosé is a blend of 80 percent Piedirosso and 20 percent Aglianico. One can also pick out broom in its bouquet, as well as a touch of violet. It will gracefully accompany a second, richer seafood course, such as *calamari* (squid) or *polipetti* (baby octopi) in tomato sauce — Vesuvian tomatoes, naturally.

As for Lacryma Christi red, also based on a similar ratio, 80 percent to 20 percent of Piedirosso to Aglianico but with a splash of Sciascinoso, it is a ruby-red wine, so impetuous and fiery that it is said to spring directly from the volcano. The better vintages can age up to ten years, developing flavors of red berries and spices, not unlike Châteauneuf-du-Pape and other GSM wines (Grenache-Syrah-Mourvèdre) of the French Rhône Valley. Red Lacryma Christi pairs well with red meat and game, veal and poultry, strong cheese, and of course Neapolitan lasagna.

The Cantina's 5 hectares (12 acres) of vine are spread out over a lava field, emplaced during the 1744–1755 eruption, with the 1822 flow terminating a short distance upslope and the longer 1906 flow grazing the estate just to the east.

Living Dangerously

Although the 1906 lava flow spared the Russo vineyard, the eruption that year was particularly disastrous and serves to underscore the many dangers that still threaten the area today.

Throughout the 1700s and 1800s, the activity of Mount Vesuvius had remained confined to the summit crater, except for a few trickles of lava down the mountain face: mild eruptions that enchanted writers and painters of the Romantic period, such as William Hamilton, Johan Dahl, Joseph Wright, and William Turner.

The 1906 eruption was quite another matter, as it escalated to the point of becoming a sub-Plinian eruption: it was similar in nature to the Plinian eruption that destroyed Herculaneum and Pompeii, but fortunately on a smaller scale.

The eruption began on April 4, 1906, with fissures opening up on the southern flank of Mount Vesuvius and sending lava flows down through the pine forests and vineyards of Boscotrecase. On April 8, the gas-rich magma proceeded to shoot out directly from the summit crater, shaking the volcano's plumbing system so severely that the summit cone began to fail, "splitting open like the petals of a flower," according to British reporters of the Associated Press who covered the event.

As it were, the entire summit of Mount Vesuvius collapsed, its altitude dropping a full 100 meters (330 ft.). Fountains of magma rose 600 meters (2,000 ft.) above the crater rim, feeding a plume of ash and gas that rose to an altitude of 13 kilometers (8 mi.).

Scoria and fine ash rained down on villages closest to the crater, and in particular on San Giuseppe Vesuviano to the east, where the roof of the church collapsed under the weight, claiming more than one hundred lives. In nearby Ottaviano, most villagers managed to flee, its five churches likewise crumbling under the ash load.

The exodus spread to the coastal villages of Ercolano, Portici, and Torre del Greco, thousands of refugees fleeing west toward Naples and east toward Castellammare di Stabia. All in all, there were two hundred casualties and thirty-four thousand refugees, and over 10,000 hectares (25,000 acres) of prime farmland covered by ash and mudflows. The latter were caused by the rising content of water vapor that was belched into the atmosphere, raining down onto the slopes and turning the blanket of

Wine cellar (Fuocomuorto Estate, Ercolano) dug into Mount Vesuvius's 1631 lava flow. The cellar was overrun by a mudflow in 1906.

ash into chocolate-colored streams that cascaded down gullies, swamped cropland, and tore through villages, most critically on the northeastern and southwestern slopes of the volcano.

The Oliviero vineyards, above Ercolano, were overrun by one of the 1906 mudflows. One would never guess it today, when visiting the estate — Azienda Fuocomuorto — as all traces of the disaster have disappeared. When I toured the premises with current owner Vincenzo Oliviero, I found no hint of the mudflow, only bedrock poking out here and there from the thin soil cover: a ropy-textured lava flow dating back to the great 1631 eruption. Vincenzo pointed to patches of gray lichen clinging to the rock: the indigenous species of *Stereocaulon vesuvianum*. As already mentioned, this lichen is one of the first life-forms to colonize fresh lava, breaking down its minerals to deliver a crumbly, clay-rich soil, well-suited to agriculture and vine growing.

Vincenzo's vineyard was established on the 1631 lava flow about a century and a half after the eruption — the date 1780 is engraved on the cellar wall — which is about the time it takes for lichen, vegetation, and humans to break down fresh rock into manageable land.

The family enterprise came to a grueling halt in 1906, when the mudflows fueled by that year's eruption raced through the vineyard, encircled the house, and cascaded into the wine cellar through the window wells, filling it from floor to ceiling and setting like concrete. The discouraged owners gave up their wine trade and scaled down their activity to simply growing a few grapes, fruit and vegetables.

Estates on volcanoes might well be struck down by eruptions, but given time, and with family tradition running strong, they end up rising back from the ashes. So it was with the Fuocomuorto estate, the stubborn commitment of its owners underscored by their very names: the founding father in 1780 was one Vincenzo Oliviero; the owner at the time of the catastrophic 1906 eruption was also named Vincenzo Oliviero; and his grandson and the present owner, who set out one hundred years later, in 2006, to shovel out the cellar, prop up the vines, and resume the wine making, was yet another Vincenzo Oliviero. A year later, in 2007, Vincenzo was pressing his first vintage from the grapes of old surviving vine and newly planted stock: Piedirosso, Aglianico, Primitivo, as well as Caprettone, the local white grape that the winemaker rehabilitated as an original varietal wine.

And so life goes on under the volcano, despite the threat of ash falls and mudflows, and the prospect of having to start anew after each eruption.

Farming the Last Lava Flow

The last eruption of Mount Vesuvius took place in 1944, in the midst of World War II, launching a lava flow that raced down its northwestern slope, destroyed the villages of Massa and San Sebastiano, and claimed thirty lives.

This is where I visited a farming estate deeply rooted in history just a couple of years before it closed: Casa Barone. From the estate's vantage point, on the northwestern flank, the volcano shows its double structure, with the ruins of the older cone — Monte Somma — wrapping around the central, more recent summit. The 1944 lava flow funneled down the valley between them, stopping short of the first villages.

Where the flow terminates, 300 meters (1,000 ft.) above sea level, lies the heartland of Vesuvius fruit farming, within the boundaries of the volcano's national park (Parco Nazionale del Vesuvio).

Eleven hectares (20 acres) in area, Casa Barone owed its name to the nobility title of its founding landowner, Catalan baron Francesco Piromallo, who established the estate in the sixteenth century. After many turns of fortune, including its dismemberment under Napoleonic rule, the estate began a second life in the 1870s. A stately farmhouse was built, Casa Barone proper, which survived the dramatic 1906 eruption, World War II, and the 1944 eruption, when lava ran dangerously close to the building. The farmhouse ultimately broke down during an earthquake in 1980, but its ruins still stand above the Bay of Naples.

The perilous location, hard work on steep slopes, and ever-decreasing profit margins marked the final years of the twentieth century, when the pressure of the open European market toppled the region's agricultural economy.

Only a quarter of Casa Barone's land was still farmed when the agronomist Giovanni Marino brought the estate back to life, keeping its traditional layout and commitment to organic agriculture. He touched up the balance of crops to move away from monoculture — dominated in the past by grapevines, then apricots — and raised instead a variety of typical Vesuvian produce: indigenous grape varieties, apricots, Pomodorino del Piennolo tomatoes (which became the estate's top-selling item), figs, oranges, berries, and walnuts.

Marino took me on a tour of the estate several years ago, in mid-May, a few weeks past the spectacular April blossoming of cherry trees that lights up the dark basaltic landscape with garlands of white flowers. We climbed up to the edge of the 1944 lava flow, broken up and leveled by farmer and machine to generate more terraces for vines and olive trees. Giovanni pointed out new olive trees planted on the upper terraces overlooking the lava flow, and venerable sorb trees, native to southern Europe and cultivated both for their hardwood — long used for building ships — and their tannic, sugar-rich fruit.[14]

We then reached the tomato vines, freshly planted in patches of basaltic cinder — the soil's granular texture draining away any excess moisture and its dark color soaking up the sun's heat to give the tomatoes their characteristic water-poor, mineral-rich pulp and thick skin. May is the month to rake up the soil in mounds at the foot of the tomato vines, and the estate's workers were busy tilling, lifting up a spray of volcanic dust in the morning sun.

Grapes were once the dominant crop on the farm, both as table fruit and for producing wine. Giovanni kept rows of vines on the estate's high-

Photographed by a squadron of B-25 bombers, the last eruption of Mount Vesuvius occurred in 1944, during World War II.

est terraces, 400 meters (1,300 ft.) above sea level. Moving away from the production of Lacryma Christi, a saturated market, he planned to place his chips instead on a varietal, highlighting the Greco-Roman Caprettone — the star of Vesuvian white wine for two thousand years.

Unfortunately, Casa Barone failed in its courageous attempt to diversify and reverted to monoculture. After the heyday of apricots, and then grapes, the estate abandoned the fruit and wine market to others, and since 2018 has refocused solely on its most successful crop, under the new banner slogan of "I Pomodori del Vulcano," the volcano's tomatoes.

A Look to the Future

In addition to economic hardship, farmers and winemakers on Vesuvius live under the constant threat of an eruption. It is possible they will get a respite over the next decades, or even centuries, as it appears that the volcano has slipped into a dormant phase.

Volcanologists assume that the last cycle of activity, begun in 1631, came to a close with the eruption of 1944. This would mean that there is little chance for the volcano to reawaken in the near future, as major activity cycles on Mount Vesuvius are separated by quiet intervals, lasting several centuries.

The downside of this respite is that whenever the volcano resumes a new cycle of activity, its reawakening will most likely be violent. The reason is that a long interval of repose allows magma underground to settle and decant, with gas-rich components separating out and accumulating at the top of the chamber. It then takes only a fresh pulse of magma to enter the magma chamber from below and act as a trigger, shaking things up and expelling the gas-rich layer at the top in the form of a violent Plinian eruption.

A rest period on the order of several millennia is probably necessary to set up an eruption as violent as the one in AD 79, so that a repeat of the Pompeii tragedy in the near future is unlikely. If an eruption does occur over the coming decades, it will be rather weak, but even a light coating of ash could be enough to ruin crops and upset the local economy.

By following seismic tremors, ground inflation, and the reactivation of fumaroles, volcanologists and city officials will probably have several weeks of warning time to enact evacuation plans before the next eruption. The strategy consists in pairing threatened townships on the volcano with villages elsewhere in Campania that would temporarily host their refugees. Eighteen townships, accounting for a total of seven hundred thousand people, fall into this "red zone," including Ercolano, Trecase, and Boscotrecase: less than 7 kilometers (4 mi.) from the summit, they lie within reach of most pyroclastic flows.

On the outskirts of this central area, the evacuation plan also defines a "yellow zone" downwind of the volcano (southeast, as trade winds blow from the northwest) that includes Pompeii. Several tens of centimeters of ash might fall in this zone during an eruption and trigger roof collapse. Up to a million people would need to be evacuated, the exact number depending on the wind's speed and direction during the crisis.

Besides the problem of evacuating and rehousing refugees, Campania would be struck economically as ash clogged water reservoirs, buried roads and railways, destroyed power lines, and ruined agricultural land and crops for many years, if not decades. But it will take more to discourage farmers from reclaiming their volcano and raising new crops out of its ashes.

Tourists at Pompeii.

Guide Section: Visiting Mount Vesuvius

One can visit Mount Vesuvius and the Bay of Naples any time of year, but the best seasons are early spring (April and May), when fruit trees are in bloom, and the beginning of fall (September and October), around harvest time.

Renting a car in Naples is inexpensive and practical, but one can manage without: a local train out of the Circumvesuviano station (next to the main train station) stops at Ercolano and Pompeii; and several tour operators offer bus excursions to the summit of the volcano, including a wine-tasting lunch at one of the major wineries.

Although Naples is an interesting city and its museum hosts most of the furniture, mosaics, and frescoes of Herculaneum and Pompeii, it is preferable to lodge in a hotel or bed-and-breakfast near the archeological sites, to be able to visit them early in the morning, before the tourist rush.

There is also a night boat (run by Siremar, now part of the company Caronte & Tourist), with private cabins at a reasonable price, that leaves Naples on Tuesday and Friday evenings for the Aeolian Islands (Stromboli, Lipari, and Vulcano) and Sicily's northern coast (Milazzo). One can thus plan a combined visit of the Bay of Naples, the Aeolian Islands, and Sicily's Mount Etna.

Geological and Archeological Tour

The archeological sites of Herculaneum and Pompeii open early in the morning (8:30 a.m.), which is the best time to visit: tourist buses have not yet overrun the sites, there is plenty of atmosphere to still soak up, and the lighting is best for photography. The same can be said about late afternoon (last admissions at 6 p.m., and the sites close at 7:30 p.m.), when the ruins are deserted and it is possible to walk the ancient streets virtually alone.

The ruins of Pompeii are widespread, and it is easy to spend a full day at the site. Plan to visit the eastern part of town, especially around the Amphitheater (the stadium), where most vineyards are located: they lie behind locked gates but are easy to photograph. In the same area, the best panoramic viewpoint of the streets and vineyards, with Mount Vesuvius in the background, is from the raised path along the southern walls of the city.

On the coastal outskirts of Ercolano, Herculaneum is smaller but should not be missed: it is much less crowded and better preserved than Pompeii—mud and pyroclastic flows did a better job of sealing off the buildings and their contents than did Pompeii's pumice—and more frescoes, mosaics, and furniture were left on site rather than shipped off to museums.

There are two other important sites that are much ignored and worth a visit: the aristocratic villa of Oplontis (in Torre Annunciata, between Herculaneum and Pompeii, down the street from the train station), that belonged to Nero's wife Poppaea; and a vineyard farm on the outskirts of Boscotrecase, known as the Antiquarium, with a courtyard full of amphorae and a stunning museum that highlights the objects found on site, including the nuts, figs, and other produce buried by the eruption.

As for Mount Vesuvius itself, two roads lead to the summit. The northwestern one is closer to Naples and carries most tourist buses and private cars. The southern one is a bit wilder and closer to Pompeii: shuttle buses regularly leave the archeological site for the volcano along this route. During the ride uphill, take note of the staging of microclimates and vegetation belts: greenhouses and orchards on the lower slopes; vineyards up to, and past the villages of Boscotrecase and Boscoreale; and an abrupt switch to woodland approximately 300 meters (1,000 ft.) above sea level, where the climate is too cold and too dry to break down the lava into fertile soil. Here, one enters Parco Nazionale del Vesuvio and its 9,000 hectares (23,500 acres) of pine forests.

Above 800 meters (2,500 ft.), upon exiting the forest, lava flows become much more conspicuous, with only broom bushes and wild flowers dotting the mineral landscape. At the parking lot, after paying a modest entrance fee, an easy footpath — 200 meters (600 ft.) of vertical rise — leads to the summit and circles the southern half of the crater rim. Guides can be hired on the spot to point out interesting details and explain the sequence of eruptions that formed the crater.

The summit cone is truncated by a deep circular pit, carved out by the cataclysmic blowout of 1631 and further modified by the 1906 eruption. The pit is now 500 meters wide and 300 meters deep (1,650 by 1,000 ft.), with rare fumaroles left over from the eruption of 1944, mostly on the southern wall.

The rim's footpath offers a spectacular view of the Bay of Naples, coastal villages, forests and vineyards, and the deep Valley of Hell (Valle dell'Inferno), which separates the central cone from the older encasing volcano of Monte Somma that frames it in horseshoe fashion. Coastal villages look tiny from above: Pompeii can barely be made out in the distance. The slope is so impressive that it provides a real sense of how much speed and momentum a pyroclastic flow can gather when rushing down the volcano's flank.

Wine-Tasting Tour

Surprisingly, despite the reputation of Lacryma Christi, vineyard tours and open-house wine tasting are less developed on Mount Vesuvius than on other volcanoes such as Mount Etna; few wineries are open to the public, and Italian is often the only language spoken. The following itinerary highlights four exceptions, including two well-run wineries where English is spoken: Cantina del Vesuvio and Azienda Sorrentino, both on the southeastern flank of the volcano, less than a half-hour drive from Pompeii. It is wise to reserve or arrange a visit ahead of time by phone or email.

Cantina del Vesuvio and its vineyards are perched above the town of Trecase and the volcano-girdling main road (Panoramica del Vesuvio).[15] Maurizio and Rossana Russo and their staff welcome individuals and groups on their terrace overlooking the sea and the island of Capri in the distance, with Mount Vesuvius towering over the estate.

A pioneer of wine tourism in the region (see pages 52–53), Maurizio manages a well-oiled (in more ways than one!) wine-tasting lunch, highlighting fresh produce from his own garden — namely tomatoes, basil, and olive oil — and his full gamut of wines: a dry *spumante*; a white, a rosé, and two red Lacryma Christi; and an apricot liqueur. All the wines, and the olive oil, can be bought in the boutique, and a stroll through the vineyard makes for some excellent shots of Mount Vesuvius, framed by vines and cypress trees.

Azienda Sorrentino is also reached off the volcano-girdling Panoramica del Vesuvio, 500 meters (1,600 ft.) up via Fruscio.[16] Azienda Sorrentino offers an equally delightful view of the Mediterranean Sea in the distance, a walk through its vineyard and olive grove, and a lunch featuring local olives, salami and provolone, fried vegetables from the garden, and Vesuvius apricot pie, accompanied by a *spumante*, as well as three or four volcanic wines, including Lacryma Christi from different terroirs (the Lapilli label comes from the highest-elevation, scoria-rich lots), the excellent Aglianico varietal Don Paolo, and apricot liqueur. With a degree in oenology and fluent in English, Maria Paola Sorrentino gracefully runs the show.

The volcano's northwestern flank, though closer to Naples, has not yet opened up to wine tourism but is worth exploring to discover the lesser-known vineyards and cherry orchards around San Sebastiano and Sant'Anastasia. One estate does sell its wine and tomato-based produce to visitors, Cantine Olivella, although make sure to call ahead to arrange your visit.[17]

Reaching Cantine Olivella is an adventure in itself (a GPS is really useful): drive up through the town of Sant'Anastasia and exit north (upslope) by following via Sant'Aniello, which turns into the narrow via Zazzera. A signpost marks the dirt road to Cantine Olivella, on the right side, past a row of old farmhouses. Owners Ciro and Angela Giordano speak only Italian (but you'll get by, as I did) and will show you their vineyard just outside their home, grading into the wilderness of the upper slopes. With an organic farming certification, Cantine Olivella sells the famous local tomatoes (Pomodorino del Piennolo), boxed for transport or jarred as pasta sauce, as well as six wine types, including red and white Lacryma Christi and two varietals: Caprettone and Catalanesca, the latter of which was granted its own appellation in 2011: Catalanesca del Monte Somma IGT (IGT stands for *Indicazione Geografica Tipica*, a recognition of the wine's restricted geographical origin and good quality but short of the coveted DOC label).

A great way to end the day and spend the night is to check in at Azienda Fuocomuorto above Ercolano, on the volcano's southern slope.[18] Set up as a bed-and-breakfast (only three rooms, so reserve in advance), it is ideally located next to the Herculaneum archeological site — a mere ten-minute drive. To reach Fuocomuorto, get off the highway at the Ercolano exit and drive upslope, cross the main road (via Benedetto Cazzolino), and head up a narrow street, past a chapel and a few houses and fields. If using a GPS, make sure to enter "contrada Croce dei Monti" and not "via Croce dei Monti" — another street altogether in downtown Ercolano.

Fuocomuorto's winery and courtyard are surrounded with vines. The cellar dug into hard lava was overrun by a mudflow during the 1906 eruption (see pages 55–56). From the B&B's terrace, there is a great view of Mount Vesuvius, towering above the estate, and of the Bay of Naples, downslope to the southwest. In the courtyard, clusters of Pomodorino del Piennolo tomatoes hang to dry, in the summer and fall. A charming host, Vincenzo Oliviero speaks only Italian, but here again, language is no barrier for wine lovers. He will open up his cellar to have you taste his excellent Lacryma Christi, including a delightful oak-barrel-aged Rosso-magma ("red magma"). Glass in hand, with the summit of Mount Vesuvius reddening in the rays of the setting sun, one can truly relax . . . until the next eruption.

CHAPTER 4

Mount Etna

Towering over Sicily, Mount Etna supports a whole range of ecosystems that vary by elevation and orientation. Between the citrus groves at sea level and the high-altitude pistachio and chestnut trees lies a vineyard belt that yields heady red wines, akin to Burgundies, and subtle whites with a floral bouquet. Their reputation has skyrocketed in the past two decades, thanks to the dedication of Sicilian, Italian, and foreign winemakers.

Mount Etna is one of the most active and best-studied volcanoes on Earth. It attracts volcanologists from across the globe and treats them with a wide array of eruption styles, ranging from peaceful gas discharge to spectacular lava fountains. Lava flows trickle out of the summit craters and occasionally stream out of fissure zones on the lower flanks, dangerously close to civilization.

The volcano itself is impressive in size, peaking 3,350 meters (11,000 ft.) above the Mediterranean. Its oblong lava shield stretches 50 kilometers (30 mi.) from north to south and 40 kilometers (25 mi.) from east to west, which corresponds to a surface area of 1,250 square kilometers (close to 500 sq. mi.).

Mount Etna has claimed few lives in the past. Although its lava flows repeatedly streak through vineyards and villages, there has always been ample time for people to move out of harm's way. The few accidents that do occur are usually confined to the summit craters, when hikers get caught in a blizzard, thunderstorm, or unexpected salvos of lava bombs.

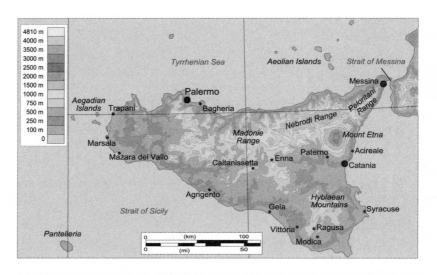

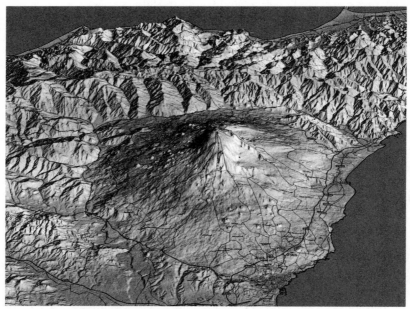

Location (top) and topographic map (bottom) of Mount Etna, Sicily.

The people of Mount Etna have developed a special relationship with their volcano: a mix of fear and respect. Its ability to send a lava flow snaking down the slope within striking distance of a village has earned it the nickname of *la vipera*: the viper. Others avoid calling the mountain by its name and refer to it only as *lui* (him). They are grateful, however, for its unfaltering fertility: Mount Etna is covered with vines, vegetable crops, and fruit and nut trees, making it Sicily's cornucopia.

Sicily's Garden of Eden

The island's climate is typically Mediterranean — hot, dry summers and mild, wet winters — but Mount Etna adds to this a vertical dimension and an extra level of complexity. The volcano intercepts and modifies wind patterns, controls precipitation, and commands a variety of climate belts, as a function of both altitude and orientation with respect to sun and wind.

Mount Etna is famous for its fruit (oranges, lemons, apples, and peaches), its nuts (pistachios, chestnuts, almonds, and hazelnuts), olives and olive oil, and of course its wine, with a reputation that goes back far in time. Vineyards were well established on the volcano by Roman times, with convenient shipping of wine amphorae out of the harbor of Catania. Wine making became the province's main asset throughout the Middle Ages and Renaissance, and reached a climax during the eighteenth and nineteenth centuries, when Etna wine was shipped across Italy, even showcased in Paris at the 1900 world's fair.

Those were the golden days. The phylloxera louse infested vine roots at the turn of the twentieth century and yielded a major blow to Etna vineyards, as did the population shift from the countryside to Catania, Messina, and other cities; a severe and crippling tax increase imposed on wine production; and increased competition from vineyards established on less rugged terrain in western Sicily. Etna's vineyards plummeted from close to 50,000 hectares (125,000 acres) in the late 1800s to a mere 3,000 hectares (7,500 acres) at the close of the twentieth century.

Within the past couple of decades, Etna wine has made a spectacular comeback, not so much in terms of area and quantity but in terms of quality. Local winemakers and newcomers are buying up old vineyards and developing the notion of volcanic terroir, focusing on indigenous grape varieties, and turning out upscale wine: hearty and spicy reds based on Nerello Mascalese, and dry, mineral whites made from Carricante. The wines owe their reputation as much to their inherent worth as to Mount Etna's growing popularity as a volcanic icon and travel destination.

Low-cost airlines and charters now place Sicily within two to three hours of most European cities, and from Catania a daily bus takes visitors and hikers up to the Sapienza parking lot and high-perched hotels, from which guided tours of the summit craters depart via cable car and jeeps.

Rental cars are a must, however, to visit Etna's countryside wine belt: several itineraries are recommended by the tourist office — such as *la strada del vino dell'Etna* — that reach approximately fifty vineyards, a dozen of which are particularly well established to offer tours and wine tasting with little or no notice. Other villages and agricultural sites promote their fruit orchards, as well as honey, chestnuts, and pistachios. All owe their quality and originality to Etna's unique terroir: a combination of volcanic soil, Mediterranean mountain climate, and deep-rooted tradition.

Europe's Most Active Volcano

Mount Etna is no ordinary volcano. Its sheer size is exceptional, and the peculiar chemistry of its lava — sodium- and potassium-rich basalt — makes it difficult to classify in the framework of plate tectonics.

The volcano straddles the tectonic boundary between the African and Eurasian plates, the former diving under the latter, and one would expect that subduction-style volcanism — waterlogged, sinking oceanic crust — would generate andesitic, explosive magma, as is the case in Indonesia, Japan, and the Andes. On the Sicilian front, however, it appears that the sinking African plate has ruptured on its way down, opening up a window through which a hot mantle plume rises unimpeded to the surface.

This unique situation yields fluid basaltic magma (technically trachybasalt, rich in sodium and potassium) that surfaces as fairly innocuous lava sputtering and fountaining above the summit craters (which act as chimneys for most of the gas discharge), and as fluid lava out of fracture zones lower down the flanks.

The unusually large size of Mount Etna is the result of its incessant activity over tens of thousands of years, plastering layers of lava and ash over basaltic foundations that were laid down as far back as five hundred thousand years ago. At the time, the site was a shallow gulf, such that early lava erupted underwater. One can view this lava today above sea level, due to the tectonic uplift over the years, in the form of pillow lavas below the castle of Aci Castello, a twenty-minute drive east of Catania: the foot-wide circular patterns are cross-sections of ancient underwater lava tubes.

After this early submarine phase, Mount Etna began to construct itself above water, as a shallow-sloped, Hawaiian-style lava shield, but then switched to shorter, stubbier lava flows over the past hundred thousand years, building its upper, steeper section.

At the 2,900-meter mark (9,600 ft.), the slope shallows off to form a vast summit plateau known as Piano del Lago: a filled-in caldera that is 2 kilometers wide (1 mi.). From this platform rise steep cones with active summit craters; these are off-limits to tourists and peak at 3,350 meters (11,000 ft.). When eruptions are threatening, even the access to the Piano del Lago platform is restricted to only groups supervised by local mountain guides. From this vantage point, visitors can admire the billowing plumes of water vapor, often tainted with brown ash, or catch an eruption in the act if they are so lucky, in the form of lava sputtering over the rim of a cone or, more rarely, blowing out as an incandescent lava fountain, accompanied by a thunderous roar.[1]

Summit eruptions are not directly involved in the shaping of the rich agricultural land below, except that they do provide a sprinkle of wind-blown ash. Instead, most of the landscaping at the base of the volcano is the work of unpredictable, eccentric eruptions that break out from fissures at low elevations, much closer to the orchards and vineyards.

Such eruptions occasionally destroy crops and threaten villages, but at a pace slow enough to let the population flee or even fight back — one of the rare places where humans lock horns with a volcano and battle for control of the land.

Playing with Fire

Unlike Santorini and Mount Vesuvius, which buried entire cities under ash and mud, Mount Etna has not harbored a cataclysmic eruption in recent times. Villages and cities have been threatened nonetheless, and the attitude over the years has switched from passive fatality, with faith and religion as the only hope, to rational intervention, with bold attempts to deflect lava flows and change the course of the future. A symbolic turning point in this switch of attitude is the 1669 eruption.

Until then, only religious processions had met lava flows, drawing their faith from a miracle in AD 252, when a flow halted before the relics of Saint Agatha, a Christian martyr whom the Romans had tortured to death the previous year and whose funeral shroud, brought to the lava front, instantly stopped the flow's progress.[2]

Fourteen centuries later, on April 1, 1669, yet another procession met a lava flow threatening Catania, but this time the priest had a better plan than simply carrying relics and chanting prayers. Diego Pappalardo, chaplain of the Order of Malta and resident of Catania, headed off with fifty men to breech the solidified bank of the advancing lava flow with pickaxes, in order to divert the flow away from the city.

The eruption had begun three weeks earlier, on March 11, 1669, with violent ground shaking and the opening of a fissure 8 kilometers (5 mi.) long, from the summit zone of the volcano down to the outskirts of Nicolosi village, only 800 meters (2,600 ft.) above sea level. Lava fountains blew from the open fissure, building two cones of incandescent scoria that grew 250 meters (750 ft.) in a matter of days — close to the size of Sunset Crater in Arizona.

From the base of the twin cones, which later became known as Monti Rossi ("red mountains"), a stream of lava collected and headed downslope to the south, engulfing several hamlets along the way, with just enough time for the villagers to flee. After overriding the hamlet of Camporotondo, the lava flow seemed to hesitate, then made a turn, influenced by the local topography, and headed toward Catania and its thirty thousand citizens.

It is then that Diego Pappalardo mounted his interception plan and met the flow with his team, wearing hides soaked with water to fight off the heat. With their pickaxes and iron bars, they took turns digging into the hot rubble that lined the advancing flow, breeching the levee and diverting the lava. They might well have succeeded had it not been for an angry crowd from the neighboring town of Paternò, west of the flow, who were fearful that the lava might head their way and dispersed Pappalardo's team with clubs and pitchforks. They justified their act with words that would become the unwritten law on the volcano for centuries to come: *che*

Recent cones on the summit plateau of Mount Etna, with the conical peak of Southeast crater on the horizon.

si lasci correre il fuoco laddove la Provvidenza l'ha destinato, or "let fire run where Providence has meant it to go."

Resuming its original course, the lava flow reached Catania on April 12 and destroyed the western half of the city, advancing 2 kilometers (over a mile) out to sea before the eruption ceased on July 15, after four months and 0.8 cubic kilometers (0.2 cu. mi.) of lava added to the landscape: about the same as Hawaii's Kilauea volcano erupts in ten years.

It would be another four centuries before the Sicilian people would dare challenge again Mount Etna's fury and break the unwritten rule of nonintervention, which hinged on the fear of dealing with the consequences if a diverted lava flow raised havoc elsewhere.

An opportunity to break the status quo came in 1983, when a small lava flow, restricted to the summit area, threatened the high-perched Sapienza parking area, cable car, and tourist shops. Because no other buildings would be affected in the vicinity if the diverted lava flow ran out of control, local officials and volcanologists devised a plan to change its course by breaching its levee with dynamite. This first attempt at controlling a lava flow was somewhat disappointing: the opening that was blasted through the levee briefly drained some lava out of the channel but sealed itself back shut in a matter of hours, and the main flow went on to destroy most of the tourist area.

THE 1991–1993 ERUPTION

The eruption that broke out near the summit of Mount Etna on December 14, 1991, and terminated on March 30, 1993, was the most important in terms of duration (472 days) and volume of lava (0.3 cubic km of lava, or 0.07 cu. mi.) since the historic eruption of 1669.

As was the case three centuries earlier, the 1991–1993 lava flow became a major threat to population centers downslope, causing the local government to react and attempt to block its progress. Despite the fact that the erupting vent was high and remote – with a fissure zone 2,200 meters (7,200 ft.) above sea level, in the headwall of Valle del Bove, the lava flow funneled down into the valley in a couple of weeks, heading straight for Zafferana Etnea, a village of seven thousand people, located 10 kilometers (6 mi.) downstream.

Building on the experience gathered ten years earlier – the breeching of a lava levee with dynamite in 1983 – Italian volcanologists, government officials, and the US Navy (which happened to be in the central Mediterranean at the time, returning from the Gulf War) mounted a twofold strategy to divert the lava flow. As a first step, in January 1992, bulldozers built a rock dam 20 meters (66 ft.) high to temporarily contain the flow.

The second act consisted in diverting the lava source as far upstream as possible. Because the flow had encased itself in a lava tunnel, which insulated it from the cooling atmosphere and kept its temperature, fluidity, and discharge rate at high levels, explosives experts positioned dynamite packs in the tunnel's roof to blast open a couple of windows; then the US Navy sent two of its Super Stallion helicopters to drop 8,000-pound concrete blocks onto the exposed flow segment, to deflect the lava out of its channel and spread it thin on the surface, where it would cool. However spectacular, the dropping of concrete blocks caused only partial obstruction, slowing down the flow just a few weeks. As for the rock dam downstream, lava flowed over the obstacle by early April.

On May 27, however, a third plan was enacted, which proved the most efficient. An artificial channel was dug alongside the flow, and the levee separating them was thinned to 3 meters (10 ft.), then blasted open with explosives. Two-thirds of the lava stream bifurcated into the spillway, allowing engineers to block the partially drained main channel with boulders.

In the end, the village of Zafferana Etnea was spared – the lava front crept to a halt a few hundred meters short of the village – but the eruption had also fortuitously waned in early June, with its discharge rate dropping from 30 cubic meters per second to only 15 (1,000 to 530 cu. ft.), before completely shutting down in March 1993.

This was a turning point, however, in the approach toward eruptions on Mount Etna, and it paved the way for other, more successful diversions of lava flows in 1992, and again in 2002, so that scientists and engineers now have the experience to at least try diverting any city-threatening lava flow in the future, on Mount Etna or elsewhere.

A Train Ride around Mount Etna

A picturesque way to discover Mount Etna and its rich history is to climb aboard the two-car diesel train that snakes around the mountain: the Ferrovia Circumetnea. The railroad was built at the end of the nineteenth century to carry freight and passengers to and from the many agricultural villages on the lower flanks of the volcano. Only passengers ride the Etna train today: a travel through space and time, aboard two brightly colored cars.

As is often the case with travel, finding the station platform can be tricky. The Circumetnea train leaves not from Catania's main station, but from an uptown location near via Etnea: Catania-Borgo. On weekdays, there is a departure every other hour or so (the line is closed on Sunday and holidays), clockwise around the volcano.

As the train leaves Catania, it crosses the rugged lava field of 1669 that destroyed the western edge of the city, now ripped open by quarries and encroached upon by high-rise buildings. Past the township of Misterbianco, the view opens up on Etna's upper slopes, studded with cinder cones that outline one of the volcano's major rift zones that runs north to south. The train makes a stop at Paternò: the village whose citizens had routed those of Catania in 1669, during the lava flow diversion attempt. Past Adrano, orange groves unfurl on the volcano's southwestern flank, followed by patches of olive trees and ground-hugging cacti studded with prickly pears. Vineyards are conspicuously absent still, given the unfavorable orientation: not enough sun in the morning, and too much sun in the afternoon.

Snaking around Etna's western flank, the train climbs to higher elevations — the land becomes rocky and barren — and crosses a series of lava flows. The 1843 lava flow in particular is worth mentioning: it claimed thirty-six lives, or nearly half of all casualties on record in the history of Mount Etna. Villagers from the nearby town of Bronte had gathered at the lava front to watch its progress and were caught by surprise when it crept over a marsh, blew apart in a steam blast, and showered bystanders with rocks.

Etna's Pistachios

As one nears the village of Bronte, fruit-bearing cacti share the rocky terrain with another type of twisted shrubbery: pistachio trees. Pistachios are Bronte's claim to fame. Although the United States and Turkey are the main pistachio producers, the Sicilian village accounts for a full 1 percent of world production, in the form of a top-of-the-line variety: Verde di Bronte DOP.[3]

Pistachios, first cultivated in the Middle East, were introduced to Sicily by Arab merchants around the ninth century. One thousand years were to pass, however, before Bronte fully embraced the culture of pistachios, in the mid-nineteenth century, turning pastures and rocky, uncultivated slopes into orchards. Not only did the lava rock turn out to be a good substrate for the tree, but Bronte's climate proved exceptional: hot in the summer and cold in the winter, with slightly more rainfall than average, brought in by westerly winds.

Besides bearing valuable nuts, pistachio trees fulfill a mechanical function: they split up the lava bedrock with their powerful, twisting roots — a property they share in common with prickly pear cacti, hence their frequent association in Bronte's lava fields. True groundbreakers, they prepare the land for less robust plants and crops down the line.

Around Bronte, pistachio trees are planted between 400 and 900 meters above sea level (1,300–3,000 ft.). The slope and the ruggedness of the terrain prohibits any harvesting by machine, and because they are hand picked, Bronte pistachios are expensive, all the more so because their harvest takes place every other year only: on even-numbered years, they are green pruned — that is, the new shoots are removed by hand — and bear no fruit. Allegedly this technique allows for the plant to rest and concentrate minerals and flavors for the following year.

The fruit grow in small bunches, and as they ripen, their soft husk turns pink. It is their emerald-green flesh, however, that makes Bronte pistachios world famous, along with their sweet, grassy aroma. Bronte pistachios have high nutritional value — they are rich in proteins, vitamins, and unsaturated fats — and are used locally in ice creams, pastries, and biscotti, as well as in Sicilian pasta and pesto, sausage, and cured meat. Nearly 80 percent of the production is shipped to mainland Italy and abroad:

THE ORANGES OF MOUNT ETNA

Along with pistachio trees, citrus trees were first introduced to Sicily during the ninth century AD by the Arabs, but it wasn't until the eighteenth and nineteenth centuries that the island became a major hub of citrus production and export. Today, citrus fruits are Sicily's dominant agricultural produce, including grapefruits, lemons, clementines, mandarins, and especially oranges. The latter account for 60 percent of Italy's production (2 percent of the world market), with international companies like Tropicana purchasing a sizable fraction of the harvest. The plains around Mount Etna in particular are the focal point of the island's orange industry and account for nearly half its output.

Sicilian oranges come in several varieties, including Navelina and Valencia, but the most valued are blood oranges, which feature a red pulp and juice, including the Moro, Sanguinello, and Tarocco cultivars. The Tarocco is even considered the best orange in Italy, and perhaps in the world, on account of its large size, low amount of pith, seedlessness, and abundant, very sweet juice. Freshly squeezed orange juice in Italy (*sprumata d'arancia*) comes mainly from Tarocco oranges.

Blood oranges as a whole have twice the amount of vitamin C as other oranges and also contain vitamin A, as well as antioxidants, including anthocyanins responsible for the juice's red pigment. Potassium and calcium content is high as well, which comes as no surprise in view of the terroir: a mix between basalt (potassium rich) and limestone (calcium rich).

Oranges do remarkably well in the plains south of Mount Etna, around Catania, and west, both in the plains and up the Simeto River valley. The climate is perfect for citrus fruit: little rainfall, as well as hot temperatures during the daytime and cold ones at night. With averages highs and lows of 24°C (75°F) and 5°C (41°F), respectively, during the spring season, the temperature swing helps concentrate sugar and anthocyanin in the fruit. The maturing and harvest season stretches from January through June.

Sicilian blood oranges are recognized by a Protected Geographical Indication (PGI): Arancia Rossa di Sicilia (Red Orange from Sicily). A number of orange groves offer visits and even lodging, in the spirit of agritourism, as do Gianferrante in Paternó (www.gianferrante.it), San Leonardello in Giarre (www.sanleonardello.it), and Le Case di Civi in Mascali (www.lecasedicivi.com). There is also a yearly festival celebrating blood oranges, La Sagra dell'Arancia Rossa, that takes place one weekend between mid-February and early April, in the perched village of Centuripe, with street events, orange *sprumata*, and orange-based recipes, including a popular blood orange and red onion salad.

Bronte pistachios are used in Swiss candied pralines and in some elaborate French foie gras.

There are over one thousand pistachio growers in Bronte who share 4,000 hectares (10,000 acres) of land and sell their harvest to a dozen companies or *aziende*. These in turn roast and package the nuts, or process them in various ways (pistachio paste, pistachio butter), and do the shipping and marketing. The harvest itself is a family tradition that takes place during late August and early September—odd years only, since in even years the trees are at rest—but regardless of the year, on the last weekend of September and the first weekend of October, Bronte hosts a pistachio fest, Sagra del Pistacchio, with food stands, concerts, and other street entertainment.

Strawberries and Wine

Resuming its loop around the volcano, the Circumetnea train climbs over a lava flow dating back to 1651 that displays a ropy, taffy-like texture (geologists call it *pahoehoe*, in reference to Hawaiian lava), then winds up the slope toward the village of Maletto, 1,000 meters (3,300 ft.) above sea level.

Maletto is known for its strawberries, another crop that does very well on the northern slope of Mount Etna. Contrary to pistachio trees, though, which colonize blocky lava flows, strawberries are confined to terraced patches of fine ash. They get most of their water from the morning dew, frequent at such high elevations, as well as from groundwater drawn from several wells.

The Maletto cultivar, introduced in the early 1900s, happens to be French: going by the name of Madame Moutot, it is fragile and very fragrant, and does best in a sunny but cool climate. It concentrates iron and calcium, which happen to be two of the top metal oxides found in basalt, and is rich in vitamin C. Besides being sold as fresh fruit in local markets, Maletto strawberries are used in ice cream, pastries, and jam. Taking its cue from nearby Bronte, Maletto also organizes its own strawberry fest, Sagra della Fragola, the second weekend in June, with street performances, a strawberry-cake parade, unusual recipes (like strawberry risotto and strawberry pizza), and everything red and Italian, including of course a red Ferrari.

Maletto is the highest point that the railroad reaches. As it begins its descent toward the Mediterranean, the next stop is the city of Randazzo: the northernmost point—or "noon" position of the clockwise loop around the volcano. It is also the halfway mark and switchover point of the single-track line: one changes trains at the station to continue eastward and the initial one backtracks westward to Catania.

From Randazzo onward, the train penetrates the northeastern sector of the volcano, made up of contrasting landscapes that again illustrate the staging of microclimates as a function of altitude.

At the base of the lava shield, at the junction with its limestone pedestal, runs the Alcantara River, lined with orange and peach groves. Above the stream, on the lowermost lava slope, between 400 and 800 meters elevation (1,300–2,600 ft.), grow the vineyards, often set on narrow flat terraces of ash and scoria held in place by lava-stone walls — a terracing pattern that underscores the contour lines of the volcano's topography. Many of the terraces date back to the heyday of wine making on Mount Etna, more than a century ago, and are now abandoned, claimed by shrubs and olive trees. Still higher up stand groves of oak, beech, and chestnut trees, and beyond that an alpine microclimate where only pine trees persist, up to 2,000 m (6,600 ft.). Above the tree line, the volcano is barren, with twisting tracks of lava flows and rows of cinder cones leading up to the summit.

One very recent lava flow emerges from the woods and snakes down to the Alcantara riverbed, crossing the railroad — which was rebuilt over its rubble — and skirting the eastern edge of Randazzo. Fresh out of graduate school, on my first trip to Mount Etna, I witnessed the tail end of this 1981 eruption as our train was stopped by a dark tongue of steaming basalt — an eruption that went down on record as one of the fastest outflows of basalt on the volcano, taking everyone by surprise.

Lava broke out on March 17, 1981, only hours after a salvo of earth tremors, at an altitude of 2,500 meters (8,250 ft.). The erupting fissure spread downslope, opening up like a zipper. From the crack, gas-rich lava fountains rose high into the air, feeding at their base a voluminous lava flow that progressed 8 kilometers (5 mi.) in only three days, branching out, surrounding and suffocating hundreds of sheep, goats, and other livestock. Fortunately, there were no human casualties, and the eruption ceased as fast as it started, on March 23, but the town of Randazzo was nearly overrun, and several orchards and vineyards on the eastern side of town were buried under the flow front.

Volcanologists are still baffled to this day by the exceptionally fast flow rates of such eccentric eruptions on Mount Etna, be it the 1981 flow or the 1669 flow that reached Catania, whereas most flows are slow trickles that take months to reach the same distance and the same volume. My own guess is that the "flash floods" result from the rise of the magma column nearly to the summit of the volcano, accompanied by the bulging of the edifice, which causes the flank to rip open at lower elevations: it is then the hydraulic weight of the magma column that pushes out the lava at such a high pressure and discharge rate.

Whatever their cause, the random location and speed of emplacement of such eruptions make it impossible to deflect their course or otherwise mitigate their destructive power: a stark reminder that agriculture and life in general on Mount Etna will always be risky and sometimes heartbreaking.

Randazzo is the northern gateway of Etna's wine belt: a crescent that wraps around the eastern half of the volcano, from 400 to 800 meters (1,300–2,600 ft.) above sea level for the most part, but with a number of lots reaching 1,200 meters (3,900 ft.), placing them among the highest vineyards in continental Europe.[4]

The quaint railroad stations of Passopisciaro (Moio), Solicchiata, and Rovitello dot the wine belt — villages that have entered the limelight since the renaissance in Etna wines. Two lava flows later (that overran the tracks in 1911 and 1923), one reaches Linguaglossa, the hub of the wine-making trade on the eastern flank, before the final descent toward the blue Mediterranean. After crossing the 1928 lava flow, which spared the city of Mascali, the train reaches Giarre and the harbor of Riposto, once the loading docks of Etna wine, and, fittingly, the end of the line.

To close the loop around Mount Etna and return to Catania, one must disembark at this point and walk a couple hundred meters down to Giarre's main station, on the Messina-Catania line, to catch a modern train back into the city.

A Brief History of Wine

Mount Etna's agricultural and viticultural vocation is deeply rooted in history. The volcanic province was first colonized by Indo-European tribes — the Sicels or Siculi who gave their name to the island — and then by the Greeks. Vines were introduced as early as the seventh century BC, and by the time the Romans took over, the reputation of Etna wine was already well established. Geographer and historian Strabo (c. 64 BC–c. AD 24) linked its quality directly to the volcanic terroir: "The fields of the Catanaeans are covered with ash-dust to a great depth. Now although the ash is an affliction at the time, it benefits the country in later times, for it renders it fertile and suited to the vine, the rest of the country not being equally productive of good wine ... just as wood-ashes nourish rue, so the ashes of Aetna, it is reasonable to suppose, have some quality that is peculiarly suited to the vine."[5]

The agricultural and wine-making techniques introduced in Roman days — such as the screw winepress, driven by a rotating beam — remained virtually unchanged through the centuries and are still in use in many historical wineries or *palmenti* on Mount Etna. Such traditional wine making is encouraged today by an active association of winemakers, I Vigneri, who perpetuate the techniques and spirit of Catania's first winemakers' guild, founded in the sixteenth century: La Maestranza dei Vigneri.

From the Renaissance onward, under a growing demand for wine, vineyards spread eastward along the coast to Acireale, Viagrande, and Trecastagni. Riposto became the main exporting harbor, shipping wine to mainland Italy, France, and Germany. One of the great qualities of Etna wine was indeed that it traveled well. Sicilian grape varieties are tannic, and wine aging in wooden casks — oak or chestnut — provides extra tannin and conservation potential.

By the mid-nineteenth century, the area planted with vine peaked at 45,000 hectares (112,500 acres): double the figure in Burgundy today. But after reaching its apex, wine making on Mount Etna collapsed, as vineyards were hard hit by the phylloxera louse in the 1880s. The tragedy was compounded by the severing of commercial ties between Italy and France, a major client, and the final blow came from the Italian government itself, which raised taxes on wine in the 1920s.

Faced with so many problems, vine growers scaled down their production and turned to lower-taxed, more profitable crops, such as citrus fruit. The acreage devoted to vines plummeted from 30,000 hectares (75,000 acres) in 1900 to fewer than 3,000 hectares (7,500 acres) in the 1960s, and vines disappeared from the upper slopes of the volcano altogether.

By 1968, measures were taken to uphold at least the quality of Etna wine: it was granted an appellation of controlled origin (DOC), which sets guidelines for geographical location, grape varieties, and yields per acre. As a result, vineyards are making a slow comeback on the volcano: they have rebounded from 3,000 to 4,000 hectares (9,900 acres) today and are reclaiming high ground above 350 meters (1,150 ft.) in a concerted effort toward better quality.

Most vineyards are small, family-owned lots that sell their wine directly to neighbors and local restaurants. The more formal wine trade — catering to regional, national, and international markets — is in the hands of a small number of companies that own larger estates and have enough capital to develop modern wineries and distribution networks. From less than a dozen at the turn of the twenty-first century, the number of these major players, and the area they control, is slowly on the rise, in step with the growing market for high-quality wine and the arrival of new investors banking on the great potential of Etna's terroir.

Etna Grape Varieties

Etna wines owe their quality and character not only to their volcanic setting and microclimate but also to a particular suite of grape varieties, raised on and adapted to the land over the centuries. Several dozen cultivars were introduced by the tribes and civilizations that settled Sicily, reaching close to forty in the early 1800s and still numbering twenty-two in the late 1970s, when Italy's Ministry of Agriculture conducted its last formal survey. When it comes down to business, however, Etna wine making relies on just half a dozen varieties, with the red grape Nerello Mascalese leading the pack.

Nerello Mascalese

Nerello Mascalese (called Nireddu in Sicilian) appears to have originated in the county of Mascali — hence its name — on the southeastern slope of Mount Etna. The variety is mentioned in the county as early as the 1600s, but it probably goes back much further in time, and its exact place of origin is unknown.

Nerello Mascalese is also grown farther east up the coast, near the Strait of Messina, where it stars in the Faro DOC appellation, and it has recently spread northwest and southwest to Palermo and Agrigento, becoming the second most abundant red grape in Sicily behind Nero d'Avola.

Nerello Mascalese is a vigorous vine, bearing cylindrical or conical grape clusters, often with a lateral wing, and dark blue to purple grapes. It is a late-ripening variety: its harvest on Mount Etna usually begins in early October, or even later in the highest-perched vineyards.

Wine produced from Nerello Mascalese has a bouquet of cherry, raspberry, and spice (licorice in particular). Its tannin, acidity, and high alcohol content (14 to 15 percent) make it an excellent candidate for aging in oak barrels. Hence, Nerello Mascalese is often compared to Pinot Noir, and its wine is similarly light colored (a light shade of ruby): although the grapes are unrelated, they are both exceptionally low in coloring pigments (anthocyanins).

While it is possible to make a good varietal wine solely from Nerello Mascalese, Sicilian winemakers like to pair it with another grape that brings mellowness and a deeper red color to the blend: a role that traditionally falls to Nerello Cappuccio.

Nerello Cappuccio

The number-two red grape on Mount Etna is five to ten times less abundant than its leading partner. Nerello Cappuccio, which also goes by the name Nerello Mantellato, has a rather obscure origin, and despite the "Nerello" name that it shares with Nerello Mascalese, the varieties are unrelated: DNA sequencing connects it instead to Tuscany-rooted Sangiovese. The grape gained a foothold in Calabria, then spread to Sicily, where it was first mentioned in the early 1800s in the vineyards of Trecastagni and Viagrande. The second part of its name — *cappuccio* (which means "hooded") or *mantellato* (which means "mantled" or "cloaked") — refers to the drooping arrangement of leaves around the grape clusters. The latter are conical or cylindrical, like those of Nerello Mascalese, and the grapes are deep blue in color.

A late-ripening grape, like its blending partner, Nerello Cappuccio brings to Etna wine a deep color and a range of polyphenols that complement quite nicely those of Nerello Mascalese — namely, a touch of vanilla (high vanillin content), red fruit flavors (especially cherry), and stewed fruit (benzenoid aromatic molecules). As a pure varietal wine, Nerello Cappuccio is colorful, fruity, and ready to drink young, similar in this respect to Merlot, but it lacks the tannin and structure to age well, hence the need for a blending partner. If one were to draw a comparison to Bordeaux wine, Nerello Cappuccio complements Nerello Mascalese in the same way that Merlot complements Cabernet Sauvignon.

Carricante

The leading white grape on Mount Etna is Carricante, which first appeared near Viagrande, on the southeastern slope of the volcano. It owes its name to its high yield of grapes (*carricante* means "loaded" in Latin). The clusters are large; the grapes slightly elliptical, yellowish green in color, with an orangey bloom.

Carricante can withstand cooler growing temperatures than the red varieties, and for this reason it is found at higher elevations, namely in the township of Milo on Etna's eastern slope, where the best vineyards are perched 900 meters (3,000 ft.) above sea level. Some lots on the northern slope even reach 1,200 meters (4,000 ft.), above the town of Passopisciaro.

Carricante yields a wine rich in malic acid. Winemakers lower this acidity by having it rest on its fine lees, promoting the transformation of malic acid into softer lactic acid. This malolactic fermentation mellows the wine but still maintains a decent level of acidity and fresh crispness, which distinguishes Carricante from other Sicilian whites.

As a young wine, the varietal shows subtle aromatic differences from one vineyard to the next: it is particularly sensitive to the terroir effect of soil and climate—a chameleon-like behavior, akin to that of Riesling in Alsace. And like Riesling, Carricante develops exotic fruit flavors over time, and sometimes a note of petrol (a naphthalene aromatic compound). Carricante is the leading grape variety of white Etna wines (Etna Bianco DOC), where it accounts for a minimum of 60 percent of the blend.

Minnella

Minnella is Etna's second native white cultivar, characterized by long pyramidal clusters and the oblong grapes responsible for its name: the word *minnella* is a diminutive of *minna*, which means "breast" in Sicilian. Minnella is an early-ripening grape that is usually harvested in mid-September, two to three weeks before other varieties. As a pure varietal wine, it is a dry, strongly aromatic white, with a bouquet of ripe fruit and licorice. It can also be set aside and blended with later-ripening grapes: a maximum of 10 percent Minnella is authorized in both white and red Etna DOC wines.

Catarratto

Also an endemic variety to the island, Catarratto has become Etna's number-one white grape (in terms of quantity), and even the number-one grape overall, as it now surpasses Nero d'Avola and accounts for a third of all vineyards in Sicily.

Like Carricante, Catarratto owes its name to its great productivity, yielding "cataracts" of grape juice. It comes in two main varieties— Catarratto Bianco Commune and Catarratto Bianco Lucide—with large conical to pyramidal clusters and grayish-green grapes that take on a pinkish bloom.

Catarratto wine is rich in sugar and alcohol. It is seldom sold as a varietal but enters many a blend in Sicily, including Marsala wine. Grapes are even shipped to northern Italy, bringing their high alcohol content to the blends of cooler climes. In the Etna Bianco DOC, Catarratto can represent up to 40 percent of the blend.

Other Grapes

There are many other varieties grown on Mount Etna, both local and imported, but they represent only a small proportion of the vineyards: Nero d'Avola, Vesparola (a local white grape), Naples's Coda di Volpe, Tuscany's Trebbiano, Muscat of Alexandria (called Zibbibo in Sicily), Greco Nero and Greco Bianco, and Inzolia from Southern Italy.

Several foreign grape varieties have also gained a foothold on Mount Etna, starting with Grenache (locally, and mistakenly, called Alicante, which is a different cultivar). Small patches of Grenache are found on the eastern and northern slopes of the volcano, especially around Lake Gurrida, near Randazzo, where the cultivar has original, pre-phylloxera rootstock. Because the lake overflows in the spring and floods the surrounding vineyards, disease-carrying larvae, including of phylloxera, are systemati-

I VIGNERI, KEEPERS OF TRADITION

The Sicilian oenologist Salvo Foti founded I Vigneri, a federation of small, family-owned vineyards, to promote traditional vine growing and wine-making practices on Mount Etna.

According to tradition, each vine has its own individual chestnut stake and is pruned into a compact shrub (*goblet* pruning). The rows are staggered, like rows of seats in a movie theater, so that each vine gets to "see" the sun (minimum shadowing effect).

The spacing between vines is set at 1.0 to 1.2 meters (3–4 ft.), which equates to 7,000 to 10,000 vines per hectare (2,800 to 4,000 vines per acre). This is about twice the density of vineyards in Bordeaux, for instance. A consequence of this high planting density is that vines compete for scarce water and yield grapes that are restricted in size, with a greater skin-to-juice ratio, and hence a greater proportion of polyphenols and other aromatic molecules concentrated in the skin.

The number of grape clusters per vine is limited as well—usually two or three—which likewise concentrates aromas derived from the soil. This is different from modern techniques, in which vines are trained and pruned to produce a dozen clusters or more, with high yields but lower quality.

Chemical treatment is limited, with only light spraying of sulfates in the spring, especially since the volcanic soil is already naturally rich in sulfur. Another tradition is to manage a shallow pit around each vine stock and fill it in the spring with about 4 kilograms (8 lb.) of cow or sheep manure to act as a natural fertilizer.

cally drowned, and for that reason Lake Gurrida's Grenache need not be grafted onto disease-resistant American rootstock.

More recently, the volcano has welcomed international stars like Merlot, Cabernet Sauvignon, and Syrah. Pinot Noir and Chardonnay also do well on the highest terraces, where they find a Burgundy-like microclimate: the cool summer nights stretch out their ripening phase over longer periods of time than at lower altitudes, which translates into better-developed aromatic molecules.

The Wines of Etna

Local and foreign grapes yield a wide range of Etna wines: red, white, rosé, and sparkling. The majority of Etna DOC wines fall in the range of €10–25, when purchased in Sicily. At the lower end of the scale (under €10) are the less fancy Sicilia IGT wines,[6] which are produced outside the Etna DOC geographical boundaries, either below the minimum cutoff altitude of 250 meters (820 ft.) above sea level for the appellation or completely

Label of an Etna red wine, produced by the winemakers' association I Vigneri, with a sketch of Nerello Mascalese vine stock, and a *palmento*—a traditional lava-stone winery—in the background.

out of zone, for instance on the western flank of the volcano. Even within the coveted Etna DOC envelope, winemakers can forfeit the appellation to avoid the stringent regulations imposed on cultivars and modern wine-making equipment and instead produce original, free-spirited wines.

Etna Rosso DOC: The Reds

To qualify as Etna Rosso DOC, a wine must come from a vineyard within the boundaries of the appellation (higher than 250 meters [820 ft.] above sea level, within a crescent encompassing the eastern half of the volcano); its yield must be lower than 60 hectoliters per hectare; its alcohol content must reach 12.5 percent (in practice, it often reaches 14 or even 15 percent); and the blend must contain at least 80 percent Nerello Mascalese, with up to 20 percent Nerello Cappuccio, and eventually a small amount (less than 10 percent) of white grapes such as Minnella or Trebbiano, as long as their aromas do not overpower those of their red partners.

Within the Etna DOC zone, there are no subdivisions indicating areas of superior quality for the red wines, such as the township appellations or

Grands Crus in Burgundy or Bordeaux. However, including on the label the precise geographical area — known in Sicily as the *contrada* — has been authorized since 2011, which makes it possible to distinguish and compare different terroirs on Mount Etna. There is also a *riserva* class of Etna reds, indicating a minimum of two years in oak barrels before bottling.

Etna reds are garnet to ruby colored, with Nerello Cappuccio providing the deeper hues. Nerello Mascalese brings to the blend powerful tannins and a bouquet of red fruit, with cherry the dominant flavor, often trending toward cherry brandy, much like a Beaujolais Morgon. Etna reds also have a lot of spice — anise, cinnamon, black pepper, and cloves — as well as earthy, smoky undertones that act as a reminder, even if the varietals are mostly responsible for them, of the volcanic nature of the soil.

Etna Rosato DOC: The Rosés

Etna rosés must fulfill the same criteria as Etna reds for geographical envelope, yield, grape varieties (at least 80 percent Nerello Mascalese), and alcohol content (over 12.5 percent). They turn out to be excellent wines, especially those composed of 100 percent Nerello Mascalese (because color is not an issue, Nerello Cappuccio can be disregarded). Dry and slightly astringent, Etna rosés give off a bouquet both floral and fruity, with strawberry the leading aroma.

Etna Bianco DOC: The Whites

White wines account for 15 percent of the production of DOC wine on Mount Etna and are restricted to the same geographical area than the reds and rosés, with a minimum alcohol content of 11.5 percent. The local grape Carricante dominates the blend (60 percent minimum), combined with Catarratto (40 percent maximum), and up to 15 percent of other white grapes, such as Minnella and Trebbiano. As a general rule, however, the greater the proportion of Carricante, the better the wine.

Etna Bianco is straw yellow in color with greenish reflections. The Carricante grape provides a bouquet of white flowers, grass, and broom — bringing to mind the fragrant shrubs on Mount Etna's higher slopes — as well as apple and lemon. Etna whites are dry and mineral, with often a touch of iodine, credited to the salty spray blowing off the Mediterranean, as most white grapes are grown on the eastern slope of the volcano, close to the coastline.

Etna Bianco Superiore DOC: The Top Whites

There is only one DOC category for Etna reds, but Etna whites are divided into two subappellations: Etna Bianco DOC and an upper tier distinguished for its superior quality — the equivalent of a Village or Premier Cru appellation in France — labeled Etna Bianco Superiore DOC. The appellation recognizes the outstanding terroir of Milo, a village perched high on the northern shoulder of the Valle del Bove, a horseshoe-shaped tectonic depression. Characterized by slightly more rainfall than the rest of Etna's wine belt, with a topographic orientation that casts the vineyards into shade in the late afternoon, and an exceptionally high elevation (up to 1,000 m, or 3,300 ft., for some lots), Milo's terroir offers the Carricante grape ideal growing conditions to produce wine comparable in quality and character to a French Chablis. Less than 40 hectares (100 acres) of vineyards around Milo are eligible for the Etna Bianco Superiore label, pending that a few other criteria are met as well, including a higher proportion of Carricante than in regular Etna Bianco (80 percent or above, versus 60 percent).

Etna Spumante DOC: Sparkling Wines

Sparkling wine is not a tradition on Mount Etna, but several estates have taken up the challenge, with mixed results. If they are produced within the boundaries of the Etna DOC wine belt and incorporate 60 percent or more Nerello Mascalese, they are entitled to the Etna Spumante DOC label. The production technique is that of other sparkling wines, with a second run of fermentation accomplished inside closed bottles (for a minimum duration of eighteen months), and an alcohol content that must reach 11 percent. For example, the Scamacca del Murgo estate, based in Santa Venerina, produces an Etna Spumante based on 100 percent Nerello Mascalese.

Most estates that produce sparkling wine, however, work outside these criteria and forfeit their rights to an Etna appellation in order to experiment and create their own brand of spumante. Antichi Vinai, for instance,

ETNA'S FRUIT BRANDIES

Many varieties of fruit brandy (*liquori*) are produced on Mount Etna's southern slope around the two historical centers of Santa Venerina and Riposto, home to the leading companies Fratelli Russo (founded in 1870) and Liquori Fichera (founded in 1871), which produce respectively 500,000 and 150,000 bottles a year.

The distilleries produce grappa from grapes (Nerello Mascalese and Nero d'Avola) that reach or exceed 40 percent alcohol, as well as a lava-red concoction of wild herbs and rose petals known as Fuoco del Vulcano (Russo) or Fuoco dell'Etna (Fichera), reaching up to 70 percent: as close as one can get to drinking hot magma.

Taking advantage of the abundance and great variety of fruit trees on Etna's lower slopes, the distilleries propose a whole gamut of fruit brandies that range anywhere between 25 percent and 35 percent alcohol: mandarin (Mandarinetto), orange (Arancello), wild strawberry (Fragolino), almond (Amaretto), prickly pear (Ficodindia), lime (Verdello), and last but not least, lemon (Limoncello), the most popular after-dinner drink in Sicily.

Another suite of fruit liqueur that includes cream, Cremoncello, has less alcohol (around 17 percent) and more of a velvety mouthfeel. In addition to the fruit listed above, flavors include Etna-grown hazelnut, chestnut, and pistachio.

an estate in Passopisciaro, proposes a sparkling blend of Nerello Mascalese and Chardonnay, while Benanti estate in Viagrande highlights Carricante (90 percent), with a dash of Nerello Mascalese and Nerello Cappuccio.

The Winemakers

Mount Etna boasts several hundred vineyards. Most are small, family-owned estates that produce wine only for private or local consumption. The commercial production and distribution of Etna wine are in the hands of the largest estates (approximately fifty of them), and their international reputation and marketing are upheld by an even smaller number: roughly two dozen estates.

In recent years, there has been a surge of interest in Etna wine and a move toward greater production and better quality, under the impetus of both local and foreign winemakers. Besides the established, historical estates and family-run businesses that are now joining cooperatives and associations (such as I Vigneri), there are a number of newcomers investing in Etna vineyards: successful winemakers from northern Italy who are bringing their expertise to Sicily, and foreigners who have fallen in love with the volcano.

Historical Estates

A good example of a historical estate is Barone de Villagrande. It was founded in 1727, and the title of baron was awarded to the family by the king of Naples (Charles VII) in recognition of the quality of the wine. Many other estates have historical roots and faded temporarily until heirs or new owners took over the business and put them back on the map, as happened with Tenuta Scilio (1815), Etna Wine (1820), Scamacca del Murgo (1850), Antichi Vinai (1877), and Benanti.

Benanti symbolizes the Renaissance of Etna vineyards. The family had owned vines for centuries, but wine production had virtually ceased when heir Giuseppe Benanti decided to revive the estate in 1988. A pharmacist by trade, convinced that Etna wines could achieve a higher standard than what was produced at the time, the scientifically minded entrepreneur launched a research program into the chemical makeup of the local soil and the phenolic molecules of Nerello Mascalese and other indigenous grapes. On the basis of these studies, new wine-making protocols were tested at the winery under the supervision of a young and talented oenologist: Salvo Foti.

The new program was a success, and Benanti wines began to garner medals in national and international wine fairs, and in 2004 clinched the coveted Tre Bicchieri ("Three Glasses") award in the leading Italian wine guide *Gambero Rosso*: a distinction awarded to only 250 wines out of 15,000 in competition that year, and a first for an Etna wine. Benanti wines have remained at the top of the chart ever since. Voted best Italian wine estate in 2007, the company has become a key reference on Mount Etna and shares its expertise with smaller estates, pressing their grapes and raising their wine.

Small Landowners

Motivated by Benanti's success story and the growing reputation of Etna wines, a number of small landowners have gone back to tending their long-neglected vines. To spearhead this revival of family lots, oenologist Salvo Foti left the Benanti estate and organized a federation of small owners. His association I Vigneri promotes using indigenous grape varieties, training them on individual chestnut stakes (*alborello* method), choosing quality over yield, limiting chemical treatments as much as possible, and restoring old lava-stone crushing pads and wineries (*palmenti*), even if it means jeopardizing a DOC classification, as modern health regulations can rarely be met in these older buildings.

The Expats in Sicily

Besides Sicilian-owned estates, a new wave of Italian winemakers — many of whom are expats — are buying up vineyards on Mount Etna, conscious of the volcano's strong potential.

New York–born Andrea Franchetti led the way: after revolutionizing Italian wine making by planting Bordeaux varieties in Tuscany, Franchetti bought several lots on the northern slope of Etna, planted Cabernet Franc and Petit Verdot next to the indigenous Nerello Mascalese, and launched into original and surprising blends that were an instant success. Franchetti then focused on a whole suite of Nerello Mascalese varietals, each from a different location and altitude on the volcano — spearheading the notion of *contrada*, the Sicilian equivalent of Cru.

Also born and raised in the United States, with Sicilian grandparents, Marco de Grazia bought vines on Etna's northern slope in 2003, above Randazzo: his Tenuta delle Terre Nere (Black Earth Estate) promotes only indigenous grapes. Both de Grazia and Franchetti won their first Etna awards in 2007: the coveted Tre Bicchieri in the *Gambero Rosso* wine guide.

The Foreigners

The fourth group of winemakers on Mount Etna are true foreigners. Pop star Mick Hucknall, lead singer of the British soul band Simply Red that topped the charts in the early 1990s, in 2001 bought a few acres of vineyards and restored an old lava-stone winery on the southern slope of Etna, producing reds and whites under the label Il Cantante, or the Singer.

Also coming from the art world — art galleries in this case — the Belgian and American art dealers Filip Kesteloot and Trente Hargrave fell in love with Etna in 2003, and set up their own winery, Terra di Trente, also on the northeastern slope, near Linguaglossa. They now own 8 hectares (20 acres) of old vines that are up to 150 years old and produce Nerello Mascalese reds and Carricante whites.

The most influential foreigner, however, might well be another Belgian expat: Frank Cornelissen. First a wine importer in his native Belgium, Cornelissen made the leap to winemaker after visiting Etna and falling in love with the land. He bought his first vine lot in 2001, sought advice from the locals, explored the land, and purchased more lots to establish an eclectic suite of vineyards distributed across the finest terroirs of Etna, centered on Passopisciaro on the northern slope. His scientific approach and uncompromising "natural wine" policy — only local yeasts, no sulfites, and no oak — make for original wines with a strong character.

The Notion of Terroir

Although they experiment with different grape varieties and wine-making techniques, Etna winemakers agree on one major point: the importance of terroir. The recognition that the volcano harbors a great diversity of microclimates and lava flows has led them to individualize and promote distinct geographical zones: *contrade* in Sicilian (*contrada* in the singular form). Etna winemakers met for the first time in 2008 to discuss the issue — a venue organized not by a native Sicilian but by New York–born Andrea Franchetti. Named Contrade dell'Etna, the event now takes place every year in the spring, with a growing number of participants: producers, journalists, and wine enthusiasts experiment the subtleties of terroir firsthand. This trend toward terroir recognition on Mount Etna led to an administrative amendment in 2011 that authorizes mention of the *contrada* on the label of Etna DOC wine.

The Southeastern Flank

Etna's southeastern flank, nearest Catania, is historically the oldest sector of the volcano planted with vines, with Viagrande and Trecastagni the two hubs of wine making. Lava flows are several hundred to several thousand years old, coarsely broken down by erosion, but winemakers favor cinder cones.

Offsetting the hot climate of the volcano's southern sector, cinder

cones provide cooler zones for the vines—by way of their extra elevation and better exposure to the sea breeze—and thus a longer ripening
period for the grapes—extra time that allows the biochemical synthesis
of more complex aromatic molecules in their juice. Hence, the Nicosia
winery placed some of its best vines on Monte Gorna; the Biondi estate,
on Monte Santo Nicola; and Benanti, on Monte Serra. The latter cone
harbored renowned vineyards during the Renaissance, with one lot in particular, Serra della Contessa, mentioned as early as 1474—today it is the
cradle of Benanti's Monte Serra wine: a strongly aromatic Etna Rosso that
has won many prizes.

The Eastern Flank

Traveling eastward, counterclockwise around the volcano, one crosses the
wide cleft of Valle del Bove by way of Zafferana Etnea, a village known for
both its wine and its honey production. The winding road climbs onto the
valley's northern slope, toward the high-perched village of Milo.

Located 750 meters (2,500 ft.) above sea level, with vineyards as high
as 1,000 meters (3,300 ft.), Milo hosts close to one hundred wineries for
a population barely over one thousand people. White grapes (predominantly Carricante) make up 20 percent of the vineyard. The site benefits
from its proximity to the sea and above-average rainfall of 1,000 millimeters per year (39 in.), twice the amount recorded elsewhere on the volcano, as well as from ample day-to-night temperature swings as a result
of the altitude. The result is an exceptional white wine that earns the special appellation Etna Bianco Superiore, as mentioned earlier. The highest vineyards are allegedly the best, with scattered century-old vine stock,
interspersed with broom bushes, as well as apple, peach, pear, cherry, and
hazelnut trees.

The Northern Flank

Swinging inland to the northeastern flank of the volcano, above the
Alcantara Valley and the villages of Piedimonte Etneo and Linguaglossa,
one enters Etna's "Gold Coast," facing north. In view of Sicily's hot climate and low rainfall, such a northern exposure is not a disadvantage as
it would be elsewhere in Europe, but an asset, with more moderate tem-

peratures and a longer growing cycle for the grapes, hence a greater range of aromatic molecules.

Etna's Gold Coast stretches from Linguaglossa — its northeastern gateway — to Randazzo at its northern extremity, via Rovitello, Passopisciaro, Solicchiata, and Montelaguardia, each village boasting a distinct terroir and wine personality.

To illustrate this diversity of *contrade*, Belgian winemaker Frank Cornelissen took me on a tour of his vineyards, dispersed across nine different sectors of the northern flank. We first stopped on a limestone hill beyond the lava apron of the volcano: Monte Colla, which Frank calls his Hermitage, in reference to the famous hill of France's Côte-du-Rhône region.[7] On Monte Colla, a southern exposure and siliceous, clay-rich soil yield an alcohol-rich, powerful wine with an aroma of red berries and cherry.

Just as exceptional is the hill of Pietramarina on the very edge of the lava shield. Here the sediment butte is surrounded by the distal lobes of a single and relatively thin lava flow. Under at most 3 meters (10 ft.) of lava, the vine's deepest roots tap the sedimentary basement, and this unique arrangement yields a rounder wine, with strawberry the dominant aroma.

Driving up the lava shield toward Passopisciaro, we reached younger flows covering up the older ones, namely the 1879 flow superimposed on the 1646 flow. Here, Frank owns old vine terraces in *contrada* Chiusa Spagnola, where he plans to build his new winery, amid vines and olive trees. Here the ground is basaltic and stony, the slope much steeper, and the wine more tannic.

Continuing upslope, past the village of Passopisciaro, we crossed the high road that circles Etna's northern flank at roughly 900 meters (3,000 ft.) elevation. It is here, in *contrada* Barbabecchi, that Cornelissen produces his prestigious Magma red wine, characterized by hints of tea, licorice, and Eastern spices.

Higher still, above 1,000 meters (3,300 ft.) in *contrada* Rampante, the soil is made up of volcanic sand derived from an old, oxidized lava flow. At this altitude, grapes have a hard time reaching full maturity, but in exchange, they yield finer, silkier wines that are the trademark of Passopisciaro vineyards as a whole, whereas wines from the neighboring town of Solicchiata, a couple of kilometers to the east, are known for their more tannic character. The extent to which this personality split is due to the texture and chemistry of the soil, rather than to altitude and microcli-

---

FRANCHETTI'S SUITE OF "TERROIR" WINES

Like Frank Cornelissen, the winemaker Andrea Franchetti has focused on the subtle differences between terroirs, marketing four different wines labeled according to the *contrada*, each from a different elevation on the mountain.

In his lowest lot (550 m; 1,800 ft.), on the edge of the lava shield where vine roots crack through the thin basalt to reach the underlying limestone, his Contrada C wine (*C* for Chiappemacine) brings out the limestone-influenced notes of cherry and raspberry, with a hint of salt from the sea breeze.

Higher up (650 m; 2,150 ft.), the wine from his Contrada P lot (*P* for Porcaria) is opulent, rounded, and well structured, with a nose of cherry, strawberry, and exotic fruit, as well as a smoky note.

At 850 meters (2,750 ft.), in his Contrada S lot (*S* for Sciara Nuova), grapes mature more slowly in the cooler air, on a scoria- and sand-covered lava flow. This setting confers finesse to the wine, and also a nose of black cherry and freshly cut hay, with a musky touch.

Finally, at an altitude of 1,000 meters (3,300 ft.), Contrada R (*R* for Rampante) is known for its oxidized basaltic sand: the wine is characterized by fine tannins, fresh acidity, and a nose of cherry, spiked with a touch of black pepper.

---

mate, is under investigation: one of Cornelissen's geologist friends is collecting soil cores for analysis.

A Very Special Vineyard

On Mount Etna, one of my favorite examples of terroir — in both the geological and the climatic sense of the word — lies in the Romeo del Castello estate, outside the village of Randazzo, on the northernmost edge of the wine belt.

Like so many other estates on the volcano, Romeo del Castello is a family affair: vines were grown here for centuries, until the late 1950s, when they were left unattended for several decades. After skipping a generation, the estate was rejuvenated by Chiara Vigo, granddaughter of the last winemaker in the family, with the help of her mother, Rosanna Romeo del Castello. Mother and daughter were convinced by Salvo Foti — the oenologist and terroir specialist who spearheads the I Vigneri association — that they were sitting on a true treasure. With the help of Foti and I Vigneri to restore the vineyard, and the experienced Benanti estate to oversee the wine-making process in its own facilities, the Romeo del Castello estate has made a triumphant comeback.

At the very base of Etna's lava shield, far from the summit, the estate had always seemed out of reach of the volcano's mood swings: its basement basalt was close to five hundred thousand years old, and not a single lava flow had reached the place ever since. Why would anything change?

A volcano like Mount Etna, however, plays by its own rules and pays little attention to probability and statistics. The March 1981 lava flow ran down the northern flank of the mountain, rolled between the villages of Randazzo and Montelaguardia, crawled over the road and railway, and headed straight for the Romeo del Castello estate. Burning through orchards and vineyards, the lava steamroller slowed down on the flats, hesitated over which direction to take, then veered to the east, out of the estate, and came to a halt in the Alcantara riverbed.

The Romeo family paid a heavy tribute to the volcano: nearly half of its 50 hectares (125 acres) of land were covered with lava, but the oldest vines were spared, including 3 hectares (roughly 8 acres) of Nerello Mascalese planted in 1937: a precious lot known as La Fruttiera, which had been waiting half a century for attention. Spared by the eruption, the old vines were restored by Chiara and Rosanna, and are now over seventy-five years old, standing proudly at the foot of the 1981 lava flow.

The lava flow is impressive on its own: a wall of black rock towering 5 meters (16 ft.) over the vines — as tall as a two-story building. It not only blocks out the north wind that sweeps down the volcano's flank but also acts as a giant radiator, soaking up the sun rays during the day and radiating the heat back onto the vines late into the night.

This remarkable lot yields a tannic wine, full of character, with a nose of wild cherries and spices: a character it draws from the superimposed oldest and youngest lava flows of Mount Etna in a vibrant tribute to the volcano's long history.

To truly enjoy such a wine — as with all great wines — one would ideally want to view the site firsthand and visit the vineyard. Mount Etna is beckoning: the following section suggests a few itineraries for volcano and wine lovers alike.

Guide Section: Visiting Mount Etna

Mount Etna is easily accessible by plane, via Catania and its international airport. Palermo, Sicily's other major city and airport, is only a couple of hours away by bus.

To visit the summit craters, one should schedule a trip between May and October, since the volcano is crowned by a snowcap the rest of the year, not to mention frequent cloud cover. If one wishes to visit only the lower slopes and vineyards, the weather is pleasant year-round: temperatures rarely dip below 10°C (50°F) along the coast during winter.

Mount Etna is most enjoyable in early spring (April and May), when citrus trees are in bloom, although the volcano's summit might be off limits. In late spring (May and June) there is a strawberry fest (Sagra della Fragola) in the town of Maletto on the second weekend of June, and in early fall (September and October) the grape harvest and a variety of events occur: Sagra del Pistacchio in Bronte on the last weekend of September and first weekend of October, the celebration of honey and terroir produce (Ottobratta) in Zafferana Etnea every Sunday in October, and Festa del Castagno (Chestnut Fest) in Trecastagni every weekend from mid-October to mid-November.

One can visit the summit of Mount Etna without need for a car: a bus leaves daily (8:15 a.m.) from Piazza Papa Giovanni XXIII, in front of the main train station, and reaches in a couple of hours the Sapienza parking lot and visiting center at 1,900 m elevation (6,300 ft.), from which cable car and jeep excursions depart for the summit. There is food and lodging on site (if one wishes to stay overnight, reserve well in advance). The bus returns to Catania at 4:15 p.m.

One can also experience a "vintage" ride on the Circumetnea train (see pages 73–78) that circles Mount Etna's lower flank. It runs several times a day from the uptown Catania Borgo station and stops in many villages of the wine belt along the way.

However, in order to access all wine estates and geological sites, a car is a must, and rental fees are very reasonable.

In terms of accommodation, Catania is an interesting city, worth visiting, but it is preferable to book a hotel or B&B in one of the many villages around the volcano (such as Nicolosi and Zafferana Etnea, the southern gateways to the volcano), or for hikers, the chalets and hotels upslope on the southern flank (Rifugio Sapienza, Hotel Corsaro) or the northern flank (Rifugio Ragabo, Rifugio Brunek).

A Geological Tour of Mount Etna

Etna's Early History

One should start Etna's geological tour by visiting Aci Castello, the seaside village 5 kilometers (3 mi.) north of Catania: get off the Catania-Messina highway at the Paesi Etnei exit and head down to the coast. Aci Castello features a medieval castle built on a lava ledge. The rocky platform displays beautiful cross-sections of pillow lava: tubes of basalt that accumulated below sea level, five hundred thousand years ago, before Etna started building its shield above water. Circling the platform to the south, one also gets to admire a pile of broken-up basalt, cemented in a fine, cream-colored matrix (known as a *breccia*) that underlies the castle: the result of water-rich eruptions and underwater landslides.

About a kilometer north along the coastal road, one reaches the neighboring village of Aci Trezza, where a cluster of lava pillars rise out of the water, a few hundred meters from shore: the islands of the Cyclopes (Isole

Ciclopi). According to legend, these are the rocks hurled by the blinded giant Polyphemus in his desperate attempt to sink Ulysses and his fleeing companions. According to science, the rock pillars came not from above but from below: magma rose through fractures in the seafloor and congealed into vertical sheets of dense rock, named *dykes*. Now set free of the encasing softer rock by erosion, the turrets stand alone: they display an elegant, pipe organ–like structure, with jointing typical of slow underground cooling and contraction. To admire these basaltic spires up close, spend the afternoon at the beach and take a swim, or rent a sea kayak or paddleboat: Aci Trezza is a popular weekend spot, with a wide range of hotels, restaurants, and shops renting beach equipment.

The Road to the Top

From the coastline, the road up the volcano first crosses orange groves, typical of the southern flank, then climbs through the vineyards of Aci San Antonio, Viagrande, Trecastagni, Pedara, and Nicolosi. To visit the vineyards and wineries, check out the wine lover's Itinerary A (pages 101–103).

Hill-sized cinder cones, built up by eccentric eruptions (a long way from the central crater), rise between neighboring villages: Monte Serra, north of Viagrande; Monte Troina, south of Pedara; and the twin peaks of Monti Rossi, towering 250 meters (825 ft.) above Nicolosi, which gave birth to the 1669 lava flow (see pages 70–71). Nicolosi is the last village on the way to the summit and a practical place to spend the night, with a number of hotels, restaurants, and supermarkets.

Exiting the village, the road crosses the 1866 lava flow: on the right-hand side, a quarry offers a great exposure of its gray trachybasalt in cross-section.[8] The road then climbs onto the 1910 lava flow, winding its way up between more cinder cones covered by pine and chestnut trees. As the grade steepens, the road snakes up the dark, vegetation-free 1983 lava flow until it finally reaches the Sapienza terrace, 1,900 meters (6,200 ft.) above sea level, host to a large parking lot, a few hotels and restaurants (Rifugio Sapienza, Hotel Corsaro), a cluster of coffee and souvenir shops, the cable-car terminal, and mountain guide headquarters for jeep rides and hiking excursions to the summit craters.

For visitors who do not have the time, desire, or budget to proceed higher, or cannot do so because of adverse weather conditions, the terrace already has a lot to offer: a panoramic view looking over Etna's southern flank out to the coastline and, on site, a row of photogenic cinder cones — Monti Silvestri, built up by a fissure eruption in 1892 — accessible by steep but relatively easy footpaths. Dark tongues of lava snake down to the parking lot and hotels from vents farther upslope that erupted in 1983 and 2003 and set several buildings ablaze. They were the focus of partially successful diversion and damming efforts, with levees of rubble raked up by bulldozers to contain the flows.

Hiking to the Summit Craters

Proceeding higher up requires proper equipment (hiking shoes, warm clothes, wind and rain gear), because Mount Etna is a true mountain with unpredictable weather changes. Besides strong winds, cold temperatures, and possible thunderstorms, one must also keep in mind that altitude leads more quickly to exhaustion — the summit is at 2,900 meters (9,700 ft.) — and greater exposure to ultraviolet rays. Sunscreen, sunglasses, and lip balm are in order. Rental shops at Sapienza and at the top of the cable-car run provide mountain equipment to those, and there are many, who arrive unprepared.

There are several options for visiting the summit area that depart from the Sapienza hub to reach the old lava-filled caldera named Piano del Lago at 2,900 meters. This high-perched plateau supports three active summit cones, higher than skyscrapers: Northeast Crater, the central Voragine/ Bocca Nuova Crater, and Southeast Crater. Access to these cones, which peak at 3,350 meters (11,000 ft.), is usually forbidden.

One can hike the whole way up to the Piano del Lago terrace and back at no cost, but the strenuous one-way, 1,000-meter (3,300 ft.) climb takes approximately three hours and expends much energy — time and energy better spent near the summit. A better option is to take the cable car (*funivia*) to its terminus at 2,500 meters (8,200 ft.), at a cost of approximately €40 round-trip, and hike the rest of the way up (one hour), along the jeep road: one gets to view old cinder cones up close, with a possible short hike to the Belvedere lookout over the Valle del Bove (at the top of a very steep drop-off, so skip this option if it is foggy or dark). Make sure not to miss the last return run of the cable car (check the schedule, around 6:30 p.m.), and plan your hike accordingly. One can also buy a one-way ascent only (€20) and return to Sapienza by hiking the whole way back down (much easier than the climb, of course). Either way, take along a flashlight, a cell phone, abundant water, and survival food and blanket.

The third option is the well-organized guided tour (approximately €70), accessible to all, including children, that includes the cable-car round-trip to 2,500 meters, and then a jeep and mountain guide to the terrace at 2,900 meters. There, you will disembark for approximately one hour at the foot of the twin Barbagallo cones (2002 eruption), with footpaths around their rims, facing the active new Southeast cone, less than a kilometer away. Its vent is dormant most of the time, but it can launch lava fountains at no notice.

Should the volcano be relatively quiet, cloud-free, and its summit not ruled off-limits, a final option is to book a private hike with an Etna mountain guide to reach the summit of the Voragine/Bocca Nuova cone, but as already mentioned, such hikes are often forbidden. Should an opportunity arise, this option is for seasoned and well-equipped hikers only.

Other Sites of Interest

Another interesting hike is to the southern rim of the Valle del Bove — the sector collapse of the volcano — with a breathtaking view over its lava-covered valley floor. From the Sapienza parking lot, drive a kilometer down the eastern road toward Zafferana Etnea. At the fork in the road — one branch proceeding to Zafferana, the other to Nicolosi — there is a pull-off and a well-signaled hiking trail: Schiena dell'Asino ("Donkey's Back"). The cobbled trail starts at an elevation of 1,800 meters (6,000 ft.) and rises steeply through a pine forest before leveling off through open meadows, out to the crest of the amphitheater at 2,000 meters (6,600 ft.), overlooking the lava field. Vertical slices of solidified magma (*dykes*) jut out from the steep slope and resemble the ruins of a lost city. Plan a little over one hour hiking each way.

Back at the trailhead, one can also follow the road down to Zafferana, at the termination of Valle del Bove. The village was threatened by the 1991–1993 lava flow (see page 72) that came to a halt a short distance from the first houses. Follow the signs through the village to "Piano dell'acqua, colata 1992," up a winding road, past vineyards, to a parking lot where a statue of the Madonna celebrates her role in stopping the lava flow just in time. One can climb atop the tongue of basalt rubble and reach a couple of stone houses upstream that were overrun by the lava. There are stands set up near the parking lot that sell honey. Zafferana is the capital of honey on Mount Etna, and it comes in a variety of flavors according to which groves the bees foraged: wildflowers or chestnut or citrus blossoms.

Finally, a word about the northern route to Etna's summit, lesser known and longer to reach than the Nicolosi-Sapienza-Zafferana southern route. It offers a completely different, more alpine experience of Mount Etna. One needs to circle the mountain, first on the highway east from Catania toward Messina. Exit at Fiumefreddo, head for Linguaglossa (follow the signs to "Etna Nord") and up the winding Mareneve road through a spectacular pine forest, past a couple of lodges at 1,400 meters (4,600 ft.) elevation—Rifugio Ragabo and Rifugio Brunek, where one can spend the night—and up to the Piano Provenzana parking lot at 1,800 meters (6,000 ft.), where a jeep and hiking trail, complete with guided excursions, continue on to the summit.

Etna Wine Roads

Many estates offer wine tasting on Mount Etna, and a few propose tours of their facilities and vineyards as well. It is wise to plan visits ahead of time and to make sure the wineries are open, even if hours are listed on their websites. English is spoken in most major estates; in more local ones, only Italian is spoken but communicating is easy after a glass of wine.

The three itineraries here highlight only a few estates according to the quality of their production, historical character, and atmosphere. Many others are worth a visit, and the reader should research their websites, starting with the downloadable catalog (in Italian) compiled by Catania's Chamber of Commerce, although it is now slightly out of date (2008) and lacks a few new wineries.[9]

Itinerary A: The Southeastern Flank (Nicosia, Benanti, Scamacca del Murgo)

Historically, this is the first sector of the volcano that focused on wine making, because of its proximity to the city of Catania. Take the first exit off the Catania-Messina A18 highway (Paesi Etnei, or "Etna Land") and proceed upslope, through orange groves, to Trecastagni. Drive through the northern outskirts of town (on via Etna), and follow signs at the roundabouts that lead to the Nicosia *azienda*, via Luigi Capuana.

NICOSIA

Nicosia was founded in the late nineteenth century: the *azienda* owns 20 hectares (50 acres) of vines on the volcano, as well as lots in other regions of Sicily.[10] Old winepresses and tools line the hallway leading to the restaurant and bar. Wine tasting and guided tours take place on weekdays. The restaurant (Terra & Sorsi) is open Thursday, Friday, and Saturday evenings, and for lunch on Sunday and holidays—reservations required—and features terroir specialties.

Nicosia's wine list comprises a number of varietals, several sparkling wines, as well as red, white, and rosé Etna DOC wines, including two *riserva* (oak-aged) versions from Nicosia's prized terroir: the Monte Gorna cinder cone, 750 meters (2,500 ft.) above sea level. The wines are well balanced and good value for the money (around €10 a bottle at the shop, €20 for the *riserva*).

BENANTI

East of Trecastagni, Viagrande is another historical center of the Etna wine trade. Here, the landmark cinder cone is Monte Serra, with a park and tropical butterfly hothouse on the western side and prized vineyards on the eastern flank—a terroir that was famous as far back as the sixteenth century. The Benanti family took over the Monte Serra estate in the late 1800s, and from 1988 onward they have played a major role in improving the quality of Etna wine (see page 89).

The Benanti winery is located on the main road north out of the village, toward Zafferana Etnea.[11] Where the road switches names from via Catania to via Garibaldi, a sign "Monte Serra" over a pink-walled corner building signals the turnoff toward the estate. Make reservations ahead of time. A guided tour of the historical *palmento*, the Monte Serra vineyard, and the wine-tasting session, cost on the order of €50 per person. The tour features a number of award-winning wines, accompanied by local specialties: cold cuts and cheeses, tomatoes and olives, nuts and pies. Wines usually include a sparkling Carricante, an Etna white (Pietramarina, the first Etna wine to receive an award), and the equally famous Etna red: Serra della Contessa.

SCAMACCA DEL MURGO

For a change of landscape and a different perspective on Mount Etna, follow the road north, out of Viagrande, to Zafferana Etnea (where one can buy honey and view the 1991–1993 lava flow; see page 72). From there, head east toward Santa Venerina.

The Scamacca del Murgo estate is located at via Zafferana 13.[12] A steep path leads up to the winery, looking downslope over orchards to the blue Mediterranean and upslope across the vineyards to Valle del Bove and the volcano.

The estate was founded in 1870 and today comprises 25 hectares (62 acres) of vine on basaltic scoria and broken-up lava flows, over one thousand years old. There are a number of wine-tasting options, including one that includes lunch at the restaurant (€35). The Murgo estate prides itself with interesting varietals and blends, but one should focus on the Etna DOC wines, especially the oak-aged *riserva*. There is also an elegant B&B on site, the Tenuta San Michele.[13]

Itinerary B: The Northeastern Flank (Scilio and Gambino)

To reach the northeastern flank of Mount Etna and the Gold Coast of its wine belt, take the Catania-Messina highway, exit at Fiumefreddo, and follow the main road to Piedimonte Etneo, then Linguaglossa. Exit the village by way of the Milo road (SP59iv) that climbs the flank of the volcano along the Galfina Valley, toward the well-signaled estates of Scilio and Gambino.

SCILIO

The Scilio estate,[14] run by Giovanni Scilio and his British-born wife, Elizabeth, is centered on a lava-stone *palmento*, built in 1815. The family-owned vineyard was relaunched in 1982 and has become a reputed winery on Mount Etna, with 20 hectares (50 acres) of old vine stock planted between 600 and 675 meters elevation (2,000–2,300 ft.) in *contrada* Valle Galfina. It proposes nearly a dozen different wines, including four Etna DOCs (starting at €12; around €20 for the excellent Orphéus red).

An engineer by trade, Giovanni Scilio loves trying out new ideas and techniques. He had a trench dug across 200 meters (660 ft.) of vertical drop to assess the many volcanic layers piled atop one another — consisting of triplets of lava bedrock, broken-down lava rubble, and topsoil—down to the clay layer that marks the base of the volcano.

Giovanni Scilio also supervised the renovation of the *palmento*, sticking to the traditional lava-stone underground construction style that offers optimal insulation against the heat and protrudes only as discrete stone terraces at the surface.

The tour of the estate is instructive and friendly (English is spoken) and includes the old *palmento*, the recent wine-making facilities, and the vine lots surrounding the buildings, at a wine-tasting cost of approximately €10 (€40 for a full meal). Giovanni and Elizabeth Scilio also built a twenty-room B&B and a restaurant on the premises, run by their daughter Luisa. The restaurant is open on weekends, and for groups on weekdays as well with advance notice.

GAMBINO

Past the Scilio estate, the Milo road leads upslope to a fork in the road: the right branch reaches the Gambino estate at 900 meters (3,000 ft.) elevation, one of the highest estates on Mount Etna.[15] It comprises 4 hectares (10 acres) of vines planted on terraces. Tasting rooms overlook the vines and the lower slopes of Mount Etna, out to the sea in the distance. A round of wine tasting costs approximately €15 (or €30 with vegetables, cold cuts, and cheeses).

Their Etna DOC wines (red and white), from their local *contrada* Petto Dragone, start at €17 a bottle. The winery also owns four hectares of Nero d'Avola and Inzolia vines in western Sicily on limestone and clay that yield two varietals and several blends.

Itinerary C: The Northern Flank (Cornelissen, I Vigneri, Franchetti)

For this itinerary along the northernmost section of Etna's Gold Coast, it is essential to set up appointments with wineries ahead of time. The access route is the same as for Itinerary B: exit the Catania-Messina highway at Fiumefreddo and follow the main belt road (SS120) to Linguaglossa. Continue on to Rovitello, Solicchiata (check out the Cave Ox pizzeria and bar and its impressive wine list), and Passopisciaro, where several wineries line the main street, including Fratelli Grasso, Antichi Vinai and Frank Cornelissen.

FRANK CORNELISSEN

Azienda Agricola Frank Cornelissen is located in Passopisciaro, opposite the church.[16] For a fee of approximately €25 (email reservations only), Cornelissen offers a tour of one of his ancient vineyards and of the winemaking premises, as well as a tasting of his MunJebel red and some of his single-vineyard wines, highlighting his natural approach to wine making, in which he avoids oak casks and even steel vats, and uses neutral-material containers for the vinification process. Plan on two to three hours for a visit, and bring a sweater for the swing through the cellar.

I VIGNERI (SALVO FOTI)

Also favoring traditional over modern wine-making methods, Salvo Foti and his association I Vigneri host wine-tasting sessions, but it is best to inquire by email where the venues are, as they occur in a number of vineyards.[17] The winemakers can also arrange a visit to an old *palmento* and offer an immersion class in traditional vine tending (a day in the field, wine tasting included). Their homemade, *palmento* wines — the grapes are crushed and the juice fermented in old, traditional lava-stone wineries — have great character.

ANDREA FRANCHETTI

A kilometer outside Passopisciaro, Andrea Franchetti set up his winery on the site of a nineteenth-century *palmento*: ask for directions when setting up your visit.[18] Some of Franchetti's grape varieties are unusual for Sicily — such as Chardonnay and Petit Verdot — but the estate is also known for its excellent Nerello Mascalese. Four terroir-oriented versions of the Etna grape each highlight a different *contrada* on the volcano (see page 94). The wine-tasting package (€25) includes a visit to the vineyard and cellar, and a sampling of a Passobianco Chardonnay from a high-perched lot on lava soil, a Passorosso Nerello Mascalese (a blend from different sites), and one of the four terroir versions.

CHAPTER 5

The Aeolian Islands

North of Sicily, the Aeolian Islands are a popular vacation destination and feature spectacular volcanoes — Vulcano and Stromboli — as well as a renowned sweet wine: Malvasia. After shrinking in the 1930s to a small number of estates in the sole island of Salina, Malvasia wine is on the rise again, regaining a foothold on Lipari, Vulcano, and Stromboli.

The Aeolian, or Lipari, Islands are a chain of volcanoes off the northern coast of Sicily. Following a major fault line, Vulcano, Lipari, and Salina make up the stem of the Y-shaped archipelago, prolonged by the western branch of Filicudi and Alicudi, and the eastern branch featuring Panarea and Stromboli.

The alignment suggests some form of plate-tectonic control, such as a classic volcanic arc in which one tectonic plate is thrusting under another, but the story is actually more complex. It appears that the southern Ionian plate ceased thrusting under Sicily a few million years ago and that its leading edge has broken off and is sinking vertically into the mantle, releasing blobs of magma in its wake that then ascend to the surface to build the string of volcanoes observed today.

Lava began to erupt on the deep seafloor one million years ago, but it took time for the volcanoes to build up above sea level. First emerged the island of Salina (430,000 years ago), then Lipari (210,000 years ago), and finally Stromboli (65,000 years ago). Both Salina and Stromboli now reach 1,000 meters (3,300 ft.) above the waters. Considering that the sea-

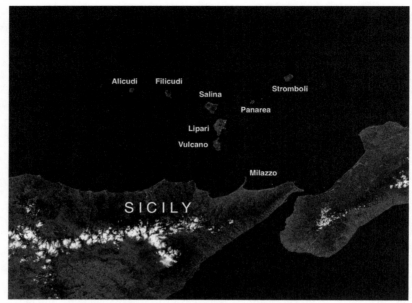

Map of the Aeolian Islands.

floor lies at a depth of 2,000 meters, the volcanoes show only their upper third; they are much more voluminous than they appear above the water. The volume of Stromboli is 200 cubic kilometers (50 cu. mi.): four times the volume of Mount Vesuvius.

Volcanoes in the Aeolian Islands display a range of eruption styles and frequencies. Several of the volcanic islands are dormant: Salina's volcanoes, for example, last erupted thirteen thousand years ago. Lipari is a historically active volcano: it last came to life during the sixth century at Monte Pilato. Vulcano has erupted on several occasions in historical times, with a last crisis from 1888 to 1890 marked by an impressive blast of large lava bombs from the central crater. As for Stromboli, it has intermittently erupted since the dawn of civilization, with sputterings of magma out of its summit craters, known as Strombolian eruptions.

Malvasia: Nectar of the Gods

The islands were named the Aeolians, because in Greek mythology they were home to Aeolus, god of the winds. Hephaestus, the ugly but clever god of the forges (later renamed Vulcan by the Romans), also took up residence in the Aeolian Islands with his team of Cyclopes, fittingly under Vulcano's steaming crater.[1] To quench the thirst of so many gods and giants, it is no wonder that this regional branch of the Greek pantheon ended up producing a divine beverage: Malvasia wine.

Malvasia is both the name of a grape variety and of the wine derived from it. Some scholars believe that the variety reached the islands in Roman times. Others believe it reached the archipelago as late as the sixteenth century, introduced by Venetian merchants. According to this theory, the variety first blossomed in Crete and was then grown in continental Greece. The very name *Malvasia* might pertain to an ancient Peloponnesian harbor, Monemvasia, used by the Venetians as a trading post during the Renaissance.

Whatever its exact origin, the Malvasia grape met with enormous success and today encompasses more than fifteen different cultivars — these are grown in the Greek islands of Paros and Syros; on the Italian mainland as far north as Friuli; in Croatia, Germany, and Austria; in Spain's La Rioja province and Portugal; in Madeira and the Canary Islands; and even in California. The cultivar in use in the Aeolian Islands, however, might well be the most ancient version.

Popular in Greek and Roman times, and again in the Renaissance, Malvasia wine from the Aeolian Islands came close to disappearing altogether when vineyards, like so many others, were struck by the phylloxera blight at the close of the nineteenth century. Nearly wiped out by the larvae, the vines were abandoned, and it was not until the 1930s that they were cared for again, under the stewardship of Nino Lo Schiavo (1905–1980).

A native of the neighboring island of Filicudi, Lo Schiavo studied to become an agricultural engineer and moved to the more central island of Salina, intent on developing and promoting local produce: timber, capers, figs, Corinthian dried raisins, and — last but not least — Malvasia wine, for which he obtained in 1973 the official appellation Malvasia delle Lipari DOC.

Around the same time, Carlo Hauner (1927–1996), an illustrious architect and designer from northern Italy, decided to retire in Salina. Besides painting — the main reason he chose the island — Hauner fell in love with wine making and established his own estate in the early 1970s. His love of innovation — he introduced the technique of temperature-controlled maceration—as well as his business and marketing savvy propelled Malvasia delle Lipari to new heights.

Thanks to its promoters, the Malvasia grape reclaimed some of its original territory, and its acreage is back to 100 hectares (250 acres), two-thirds of which are in Salina, and the balance across the other islands. Other varieties, including red grapes, are also grown in the archipelago, yielding a limited amount of interesting regional wines (appellations Salina IGT and Sicilia IGT).[2]

As for the famous Malvasia delle Lipari DOC, which must contain at least 95 percent Malvasia and up to 5 percent Black Corinth grapes, it comes in three versions. First, there is a relatively dry white (*naturale*), with a golden color, a bouquet of white flowers, a raisin-dominant aroma, hints of apricot and peach, and the salty, iodine touch from the sea mist blowing over the vineyards.

The second version is the popular *passito*: the grapes are harvested on the late side, overripe, and sun dried on reed mats for seven to twenty days. The grapes are then pressed, the maceration temperature controlled, and the wine aged in steel vats or oak casks for a period of several months. The resulting amber-colored *passito* usually fetches up to 14 percent alcohol, with a small amount of residual sugar (130 grams per liter), and aromas of figs, dates, and candied orange. The salty touch comes through as well, sometimes stronger than in the *naturale* version. It is mostly served as a dessert wine, chilled between 8 and 10°C (46°F–50°F), and can age up to several years.

The third version, *liquoroso*, is produced in the same way as *passito*, with similar scents and aromas, but fortified with extra alcohol that brings its content up to 20 percent. It is served as an after dinner digestif.

The Aeolian Islands' wine industry is split among less than a hundred vine growers, each estate averaging 1 hectare (2.5 acres) in area. The production and marketing of the wine, however, are in the hands of fewer than a dozen larger estates that buy the harvest of smaller owners. Most are established on the island of Salina, but two other islands are worth mentioning: they are famous for their volcanoes, and wine making is staging a comeback there: Vulcano and Stromboli.

Vulcano: Vines on a Time Bomb

Vulcano is the nearest of the Aeolian Islands: a ninety-minute cruise from Sicily's Milazzo harbor, or a mere fifty minutes by hydrofoil. The island, 8 by 4 kilometers in size (5 by 2.5 mi.), consists of three overlapping volcanoes. The southernmost one was first to emerge out of the waters, roughly a hundred thousand years ago, before collapsing to form a wide caldera, filled in and transformed by ash falls into a level plateau, named Vulcano Piano. This remote part of the island is where most vineyards are planted today.

The central part of the island was built up by a second volcanic cycle, roughly fifteen thousand years ago: it too culminated with explosive eruptions and the collapse of a caldera 3 kilometers (2 mi.) wide, its western rim still visible as a steep arcuate ridge framing the central depression.

Magma continued to rise in the northern half of the caldera, building a new cone about five thousand years ago — the Fossa volcano — that now towers 400 meters (1,300 ft.) above the coastline and is still dangerously active today.

Finally, a third volcanic complex emerged out of the waters two thousand years ago, to the north, spewing enough lava and ash to build an isthmus of black sand that connected it to the main island in the sixteenth century: Vulcanello ("little Vulcano").

The large Fossa volcano, towering over the harbor, is the island's main attraction. Wisps of sulfurous vapor unfurl along the crater rim, giving off a mixed impression of tranquility and deep tension, especially with the knowledge that the last eruption, in 1888–1890, was particularly violent, hurling rocks the size of cars up on its rim, and showering the harbor with fist-sized rock bombs that caused the few villagers to flee. These impressive blasts of steam and rock were christened Vulcanian eruptions by geologists at the time, who named different types of eruptive behavior after their most representative volcano. In essence, it was a phreatomagmatic eruption, in which steam produced by the underlying magma blasts out the rubble that plugs up a crater.

Although it has not stirred back to life since 1890, Vulcano's Fossa is far from extinct. A look at its historical record shows that it erupted a dozen times over the past five hundred years. Considering the average interval of forty years between eruptions, it is surprising that the volcano has slum-

Vulcano's main crater, La Fossa, which last erupted in 1890.

bered into such a long hiatus — over a century — since its last crisis. One might even fear that the unusual delay is allowing gases to build up under the crater floor, such that the next eruption could be particularly violent. In any case, Vulcano's Fossa is extensively monitored with tiltmeters, seismometers, and frequent analyses of the temperature and composition of its fumaroles.

Judging from the carefree lifestyle at the foot of the volcano, and the expansion of summer villas and tourist shops around Vulcano's quaint little harbor, the memory of the last eruption seems to have faded, especially since it caused no casualties. But those were different times, when only a dozen houses stood under the volcano. Today, the population of the sprawling village reaches ten thousand vacationers in the summer. One can only hope that for its next eruption, the volcano won't blow its top in August.

During the winter, the population of the hamlet shrinks to 250 souls (700 for the entire island), as there are few activities besides tourism. The rough terrain and lack of water support only subsistence farming and a few vines and olive trees. There was one attempt in the past, however, to exploit Vulcano's sulfur. In 1860, the Scotsman Lord James Stevenson bought the entire island, built a house at the foot of the volcano, and reopened an old quarry: the Faraglione, a yellow-tinted knoll overlooking the harbor. The eccentric lord even planted vines and exported some of his wine back to his native Glasgow.

Vulcano's small vineyard met the needs of the few islanders, until the phylloxera blight broke out in the late 1880s, followed by the volcano's brutal reawakening. Vine growers threw in the towel, and James Stevenson sold off his land at a loss, including a large parcel to Giovanni Conti, a local farmer. Giovanni's son Riccardo and daughter-in-law Isabella had the bright idea to build a hotel in the early 1900s, promoting the thermal waters of the volcano, attracting celebrities, and launching Vulcano into the era of tourism. The hotel still exists (Hotel Conti) on the isthmus of black sand leading to Vulcanello.

During the closing decades of the twentieth century, the rise of tourism brought another blow to vine growing on the island, by providing more lucrative jobs with a lighter workload. Today, however, there is renewed interest in making wine on Vulcano. At the foot of the Fossa, between the harbor and Hotel Garden, a couple of acres of Malvasia were planted recently. Most vines, however, are located away from the harbor's expensive real estate, on the southern plateau of Vulcano Piano. Red grapes are usually favored here, as the conditions for Malvasia are less favorable than in the neighboring islands of Salina and Lipari. Hence, the 10 hectares (25 acres) planted at Vulcano Piano in the year 2000 by Salina's Carlo Hauner estate are the red cultivars Alicante (Grenache), Nero d'Avola, Nocera, and more recently Sangiovese and Corinto Nero (Black Corinth). The Hauner estate named the resulting wine Hierà, in tribute to the island's name in ancient Greece.

There is very little rainfall on the island, and the soil is acidic: wild plants on Vulcano Piano are limited to broom and other shrubbery that manage to break into the sulfurous ground. Here, it isn't mildew that can ruin the grapes; the threat comes from thirsty goats and birds. Raised in such a harsh environment, Vulcano wine is nothing short of a miracle. It has a purple color, a bouquet of raspberry and cherry brandy, with earthy tones, a touch of tobacco, and even a wisp of sulfur from the volcano itself.

Lipari: The Central Island

Lipari is the largest of the Aeolian Islands, separated from Vulcano by a shallow strait. It is an interesting island from a geological perspective, with recent volcanic domes, pumice quarries, and beautiful flows of glassy obsidian (Forge Vecchia, above the village of Canneto, and Roche Rosse, north of the island). The latest eruption took place in the sixth century AD, and it certainly won't be the last.

Lipari was covered with vineyards in the past, as attested by the many abandoned terraces, but after the phylloxera blight and the rural exodus in the 1930s, estates were abandoned, and few have since been reclaimed. Only recently did the growing reputation of Malvasia and new interest in Sicilian varietals put Lipari wine back on the map.

A good place to visit is Tenuta di Castellaro, above the village of Quattropani on the road to the kaolin quarries. The 30-hectare estate (75 acres) is part of the I Vigneri consortium of traditional winemakers, headed by Etna oenologist Salvo Foti. It produces a dry white, blending Malvasia and Carricante (Bianco Pomice, or "White Pumice"); a rosé from the same grapes (Rosa Caolino, or "Kaolin Rosé"); a red wine based on Nero d'Avola and Black Corinth (Nero Ossidiana, or "Red Obsidian"); and, of course, the star wine, Malvasia delle Lipari.

Tenuta di Castellaro is open for wine tasting and guided tours that start an hour and a half before sunset (or eventually at other times for sizable groups, reserve in advance).[3] The tasting costs €25 per person for up to seven wines (€35, including the visit to the winery), with a view of the sun setting over the western islands of Filicudi and Alicudi.

Stromboli: Fire and Wine

Stromboli is the farthest from Sicily of the Aeolian Islands: a five-hour cruise by ship from Sicily's Milazzo harbor, but less than two hours on a nonstop hydrofoil run. Puffs of ash rise intermittently from its summit crater, and at night the red glow of the projections are visible from far at sea.

Stromboli has been in a constant state of eruption for the past 1,500 years. The intermittent spray of glowing magma at the summit can be viewed up close and is seldom dangerous, such that the island has become a major tourist attraction, especially in the summer (a season to avoid, if possible). Stromboli residents have managed to preserve the beauty and intimacy of their island, and excursions to the summit are carefully and efficiently managed by local guides.

Stromboli's original sputtering behavior — known as Strombolian activity — results from the confinement of a fluid column of magma inside a narrow conduit, or "chimney." [4] Gas bubbles ascend through the magma column, catching up and coalescing with one another. When they reach the surface, every few minutes on average, they burst and blow shreds of magma skyward.

Stromboli's plumbing system branches off at the summit into half a dozen vents, each of which has its own geometry and supply line, so that eruptions alternate randomly from one vent to another. The show is all the more spectacular from its place in a horseshoe depression facing the sea, below the summit ridge, so that observers on the crest have a good, downward-looking view of the vents and are usually out of reach of the projections. The glowing rocks fall back around the vents; some bounce and roll down to the sea along a steep landslide scar, known as Sciara del Fuoco, or "the way of the fire."

The steep slope is the pathway followed by the rare lava flows that spill out of the top craters when there is a surge in the magma supply. Such spectacular lava flows that reach the sea and set off billows of white steam occur once or twice a decade on average: the latest ones occurred in 1985, 2002–2003, 2007, and 2014.

Occasionally, explosions at the summit increase in violence, threatening observers, bombarding the crest line, or even propelling bombs onto the lower slopes above the village. Some of the glowing bombs can set off fires if they reach the vegetation cover, as was the case in August 1998, in December 2002 when the population was temporarily evacuated, and again in 2010.

The most violent paroxysms, when the billowing cloud of ash rises several kilometers above the summit, can even threaten the seaside villages, as happened in 1930, when a glowing avalanche raced down a gully, killing four islanders, and bombs fell on houses, claiming a fifth life and wounding many. The toll remained low because of the scarce population at the time. Today, a similar eruption during summer, when the tourist population reaches ten thousand, would have dire consequences.

The 1930 eruption upset the economy of the island. Until then, agriculture had been successful on Stromboli. Olive trees and caper bushes flourished on the volcano's northern half, and grapevines on the southern half. Islanders were largely self-reliant, fishing and growing fruit and vege-

tables in the fertile volcanic soil. As told to me in jest by the late Dome-
nico Russo (1924–2017), whose daughter Maddalena runs the lovely hotel
La Sirenetta, "You could plant a mere stick and it would start to grow."
Apricots were particularly successful, as were wild prunes, named *susine*.

Every family had its winepress. Malvasia grapes were sun dried on reed
mats during the daytime and covered at night to protect them in case of
rain. Other cultivars were grown as well, notably Minutidda — the local
name for Black Corinth.

On the heels of the phylloxera blight that struck the vineyards at the
dawn of the twentieth century, the 1930 eruption was the last straw for
the islanders and caused a major exodus: many migrated to Australia and
settled around Melbourne.

For those who stayed, tending the vines became much less interesting
than embracing the easier, more lucrative jobs provided by tourism, especially because the island was outside the recognized geographical boundaries of the Malvasia delle Lipari DOC — an appellation that was indis-
pensable to making local wine production a commercial success.

Fortunately, proud of their history and tapping the experience of their
elders, Stromboli's younger generation is returning to wine making, often
as a second job or hobby, in addition to guiding volcano tours, or running
shops and restaurants. Vines remain scarce, but the area under culture now
reaches a few hectares, dispersed in small lots, with Nero d'Avola, Nerello
Mascalese, and Syrah the leading cultivars. Nested in a dale under the Volcano Observatory, upslope of the village, pergola-trained vines also mark
the long-awaited return of Malvasia to Stromboli.

Salina: The Hub of Malvasia

Occupying a central position in the archipelago, Salina is the second
largest island after Lipari, and is made up of two volcanic cones that
earned it the name of Didyme ("the Twins") in ancient times.[5]

Salina's twin volcanoes are impressive in size — Monte Fossa delle Felci
reaches 962 meters (3,178 ft.), which makes it slightly taller than Stromboli — but neither has erupted for thirteen thousand years. In the saddle
between them (Valdichiesa, or "valley of the church"), more humidity is
trapped than elsewhere on the island, including frequent showers in the
spring and fall. The microclimate has hastened the decomposition of lava

PANTELLERIA

Pantelleria is not one of the Aeolian Islands—it lies on the opposite side of Sicily, in the straits separating it from Tunisia. One can fly to Pantelleria from Palermo, or take a hydrofoil from Mazara del Vallo or a ferry from Trapani. The oblong island is 13 by 8 kilometers (8 by 5 mi.) and consists of two ancient calderas—nested one inside the other—and a cluster of more recent volcanoes to the northwest whose last eruptions occurred less than ten thousand years ago. The lavas are silica rich, with a high sodium content: peralkaline rhyolites named *pantellerites*.

With only 7,500 inhabitants year-round, Pantelleria is a peaceful island that attracts celebrities: Sting and Madonna are among the most famous vacationers, and the French actress Carole Bouquet owns a vineyard (Sangue d'Oro label). The traditional cultivar is Muscat of Alexandria (locally called Zibbibo).

As is the case with Malvasia in the Aeolian Island, Pantelleria's Zibbibo can yield a dry white, but the sweet version is more popular, made with grapes harvested late and dried on reed mats and bleached terraces. The end result is Passito di Pantelleria, aged in stainless steel or oak casks. It boasts a cool acidity, well balanced by the sweetness of its residual sugar, and a bouquet of apricot, mandarin, and blood orange.

The total area of Pantelleria's vineyards has stabilized around 1,400 hectares (3,500 acres): 90 percent of that is dedicated to Zibbibo, and the balance to Catarratto, Inzolia, and Nero d'Avola. The vines are cultivated on terraces, hug the ground, and are protected from the strong winds by drystone walls. Naturally, the substrate is volcanic: mostly pumice, rich in silica and sodium, and a small proportion of clay.

One can taste the Passito di Pantelleria in the island's restaurants and various wineries, for example at Cantina Valenza, an old monastery in the southern foothills; or Azienda Vinicola Minardi and Donnafugata on the northern shore.

into fertile soil and supports luxuriant vegetation, principally made up of cane and ferns, as well as eucalyptus, chestnut, and oak trees.

Hence, agriculture is more developed on Salina than elsewhere in the archipelago: approximately 25 percent of the island's area is cultivated today, as opposed to less than 5 percent on Vulcano and Stromboli. Grapevines were long a big part of the picture, but as was the case on other islands, they were badly hit by the phylloxera blight. Collapsing to a mere 12 hectares (30 acres) in the 1970s, Salina's vineyard has bounced back to over 100 hectares (250 acres) today, three-quarters of which are classified under the coveted Malvasia delle Lipari DOC appellation.

Most of Salina's vines are planted on the northern coast of the island, around the township of Malfa, which is where the cleft between the two volcanoes leads inland to Valdichiesa. Vines are planted on both sides of this saddle, up to an altitude of 350 meters (1,150 ft.). Not only is the area more humid than elsewhere — with seawater spray held responsible for the wine's touch of iodine and salt — the slopes also are less steep than in other parts of the island, making it easier to work in the field.

Besides the star grape Malvasia, Salina winemakers also cultivate other Sicilian varieties, such as Catarratto and Grillo, that yield local dry whites (Salina Bianco IGT), as well as Nerello Mascalese, Nero d'Avola and Black Corinth that make up the local reds (Salina Rosso IGT).

Guide Section: Visiting the Aeolian Islands

For wine and volcanoes, the three islands not to be missed are Vulcano, Salina, and Stromboli. Lipari — the central and largest island — is also worth a visit. At the tip of the archipelago's western branch, off the beaten path, Filicudi and Alicudi attract visitors seeking the quieter, wilder aspects of the Aeolians. As for Panarea, between Lipari and Stromboli, it is known for its nightlife and jet set.

The Aeolian Islands should be avoided in July and August, when they swarm with vacationers and tourists. Spring is my favorite season, with blooming oleanders, bougainvillea, and broom. Fall comes next, with great lighting for photography.

The Aeolian Islands are easy to reach from the airports of Palermo and

Catania. Buses link the airports directly to the harbor of Milazzo, two hours away. From there, ferries, boats, and hydrofoils (*aliscafi*) depart for the islands several times a day. The faster hydrofoils reach Vulcano in fifty minutes and Stromboli in two hours. The less frequent ferries take twice as long, but their open decks allow travelers to admire the scenery and take photos. A nighttime ferry also links Naples to the Aeolian Islands and Sicily twice a week, which makes it convenient to visit Mount Vesuvius, the Aeolian volcanoes, and Mount Etna on the same trip.

There is no need for a rental car on location, as most sites are within hiking distance, except on the larger islands like Lipari and Salina, but cabs and buses suffice.

Visiting Vulcano

The island of Vulcano is famous for its volcano — La Fossa — that towers over the harbor. Climbing to the top crater takes little over an hour by way of a well-kept path that starts half a mile outside the village. The trail is quite steep in places, with a total vertical climb of 400 meters (1,300 ft.), and there is no shade: it is important to carry water, as well as a wind-breaker, since it is often windy at the top.

The view from the crest is breathtaking, overlooking nearby Lipari and the other islands in the distance, and into the dormant crater that is 100 meters deep (330 ft.). One can circle the rim and contemplate the many fumaroles that precipitate yellow needles of sulfur around their vents. There is no point in hiking down to the very bottom of the crater, which is a level arena of volcanic clay, with scattered boulders.

Another point of interest on Vulcano is the mud baths, near the harbor. The clay-rich pool was created during a drilling attempt in the 1970s to harness geothermal energy, but the hydrothermal water was not hot enough — except for the bathers who now enjoy the sulfur-rich springs (leave silver jewelry at the hotel, or it will tarnish in reaction with the sulfur). Just outside the harbor area, notice the lot of Malvasia vines along the path leading to the Hotel Garden.

Most vines are located on the southern half of the island, near the hamlet of Vulcano Piano. It is a long and steep way by foot, at 6 kilometers (4 mi.), but three buses a day leave from the harbor (the last bus returns from Vulcano Piano at 5 p.m.). The road curls around the back of the

Fossa volcano, then ascends to the plateau. One can spot small vine lots around the hamlet of Vulcano Piano, as well as farther south, where the road climbs toward Gelso, located at the tip of the island. There is a small restaurant at Vulcano Piano that serves pasta and local wine, and one can spend the night with a reservation. Also by appointment, it is possible in Gelso to meet one of the few winemakers who produces Malvasia on Vulcano (if she's on site): Paola Lantieri, whose 5 hectares (12 acres) of vines overlook the sea.[6]

Visiting Stromboli

Besides its charming and unspoiled village, Stromboli's main attraction is the volcano. If the summit is clear of clouds, there are two ways to view the eruptions. For seasoned hikers, several companies (including Magmatrek, located next to the church) offer well-organized climbs to the top of the volcano, starting around 5 p.m. (earlier hikes are not encouraged given the daytime heat, and because eruptions are most spectacular at nightfall, when the lava bombs glow red). The 900-meter (3,000 ft.) vertical ascent is tough and requires good hiking shoes, which can be rented or purchased on site, as well as a windbreaker for the summit, water and food, and a flashlight for the return leg at night. Photographers should bring along a tripod and a shutter-release cable for low-light photography. Hikers who encounter problems during the climb and can't keep up the pace are encouraged at midslope to give up the challenge and are accompanied back down to the village. Because several groups follow each other at short intervals, the time spent on the crater rim is an hour or less, which makes it possible to witness an average of half a dozen bursts of activity. The round-trip hike lasts five to six hours: groups are back at the village around midnight. Solo hikes to the top are not allowed, for safety and commercial reasons.

The second, less physical option to view the eruptions is from the northern tip of the island: the Semaforo or Osservatorio, which is in fact an open-terrace pizzeria. The hike on a cobbled mule path takes about an hour, rising 110 meters (350 ft.), but offers little shade: one should start late in the afternoon and carry water, as well as a flashlight for the return leg. The Osservatorio pizzeria offers a great view from below of eruptions on the northern edge of the crater, a kilometer distant as the crow flies.

The outdoor tables fill up for the evening fireworks display, and a camera mounted on a tripod with a telephoto lens can capture excellent imagery, with the added advantage of fresh pizzas, wine, and beer.

Hikers can also continue from the Osservatorio along the cobbled path that zigzags up the slope and after forty-five minutes reach the edge of the Sciara del Fuoco: the steep landslide scar along which incandescent scoria bounce and roll from the summit down to the sea. A couple of zigzags higher, the path hits the Sciara del Fuoco a second time at 300 meters (1,000 ft.), with spots in the bushes to settle down and watch the eruptions (again, a flashlight is mandatory for the return trip). A second observation spot can be reached another twenty minutes up the volcano, at around 400 meters elevation (1,300 ft.). It is not advised to hike up any higher without an official guide.

Vineyards are more difficult to spot than eruptions on Stromboli. There is a small lot under the church plaza visible from the terrace overlooking the sea, as well as an occasional lot, and picturesque orchards, along the narrow streets that lead down to the coastline from the main road linking San Vincenzo to the hamlet of San Bartolo. On the upslope side of the road, between the volcano observatory, perched on a hill, and the cemetery on an opposite knoll, more vines are nested in a small valley. Take the path to the cemetery to get a view, and check out the caper bushes along the way.

Visiting Salina

Salina is an island for hikers and nature lovers: its landscapes are featured in several films, most notably Nanni Moretti's *Caro diaro* (*Dear Diary*, 1993) and Michael Radford's *Il postino* (*The Postman*, 1994) — the latter recounting the exile in Salina of the Chilean poet Pablo Neruda and his friendship with the island's illiterate postman.

Salina's twin volcanoes are dormant at present, but there are some beautiful cross-sections of lava and ash in the road cuts between Santa Marina harbor and Malfa. The rain trap created by the saddle between the volcanoes makes Salina the most luxuriant island of the archipelago, with a rich bird life. It is also the center of Malvasia wine making: most wineries are located in Malfa on the north coast, including Francesco Fenech, Antonio Caravaglio, Salvatore d'Amico and Daniela Virgona, to name but a few. Make an appointment by email or phone, or try your luck simply showing up. Renting a small car or scooter can be useful on Salina; buses also leave Santa Marina harbor for Malfa every hour or so.

The bus for Leni and Rinella passes through the lush valley of Valdichiesa, where many vineyards nestle. Another picturesque site, closer to Santa Marina harbor, is Cape Faro and its lighthouse that towers over a sea of vines (6 hectares, or 15 acres), planted by the Sicilian estate Tasca d'Almerita around a fancy hotel and restaurant.[7]

On the opposite side of the island, 3 kilometers (2 mi.) south of Santa Marina, do not miss Carlo Hauner's winery on the road leading to the village of Lingua and its salt marshes. Carlo Hauner was instrumental in reviving Malvasia wine making on the island, and the estate's tasting room and wine shop are well worth the detour.[8]

CHAPTER 6

France's Hidden Volcanoes

France has its own volcanic province: Auvergne. Its most recent rift zone runs through the city of Clermont-Ferrand and the department of Puy-de-Dôme and has long been a hot spot for agriculture, orchards, and vineyards. Today, the Gamay-based appellation Côtes-d'Auvergne is making a comeback, rooted in basalt and pumice flows that enhance the grape's peppery flavor.

One does not think of France as a volcanic country, but its geological history is rich in tectonic and volcanic upheavals that have left their mark in the landscape and add an extra dimension to the variety of soils involved in agriculture and wine making.

France lies at the heart of a tectonic network that was active many times in the past, when plate boundaries framed it north and south. Thrusting of the oceanic crust under Brittany resulted in voluminous eruptions, as early as six hundred million years ago, with well-preserved lavas still visible along its coast. Much later, during the plate tectonic collision that brought together Europe and North America, around three hundred million years ago, thrusting skyward the Appalachian and Hercynian mountains, more eruptions broke out in the backcountry, piling up thick blankets of volcanic ash on the Riviera, between Cannes and Toulon — today, the beautiful red hills of the Esterel massif.

After a long cease-fire, during which dinosaurs roamed undisturbed in the Jurassic and Cretaceous periods, a new episode of rifting and volca-

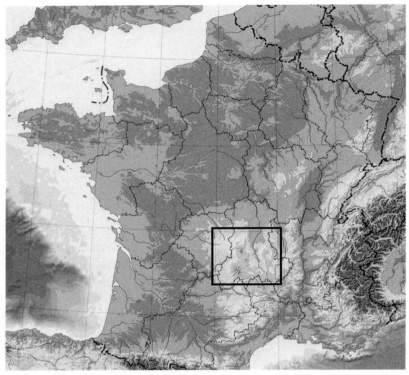

Location of major volcanic centers in Auvergne, France.

nism ripped through France as Europe, Africa, and North America drifted apart, swinging Spain and Italy into their present positions. Mountain building took place in the pinched border zones, creating the Pyrenees and the Alps, and crustal thinning and collapse occurred in the backcountry, between thirty-five million and twenty-five million years ago, opening up rift valleys trending north-south in Alsace (Rhine Graben), east of Burgundy (Bresse Graben), and Auvergne (Limagne Graben).[1]

Rift Zones in France

Rifting ceased before it managed to split France apart and resulted in few eruptions in the northern rift branches — the Kaiserstuhl volcano in the Rhine Valley is one exception — but faults are still active today, dropping the valley floors down under their load of sediment and uplifting the rift shoulders. This led to the formation of the Vosges and Black Forest mountain ranges, west and east of the Rhine Valley, and of Burgundy's Côte d'Or and Beaujolais hills.

In the Limagne rift to the southwest, volcanism was much more widespread. At the bottom of the valley, at the site of the future city of Clermont-Ferrand, lakes and swamps were shaken by explosions, starting twenty-eight million years ago, as magma made its way to the surface. Shattered lacustrine limestone mixed with basalt shrapnel and sprayed far and wide, creating a salt-and-pepper layer that would be planted with vines millions of years later.

Volcanic activity then migrated farther south, giving birth to a giant volcano, Mont du Cantal, which probably grew taller than Mount Etna stands today, peaking approximately 4,000 meters (13,000 ft.) above sea level. Construction of the lava shield began twelve million years ago and reached its apex four million years later, before shutting off.

Worn down by erosion to half its initial height, the old Cantal volcano now spreads its ridges in a starfish pattern, its arms separated by deep valleys. The steep slopes are better suited to cattle raising and cheese production (namely Cantal AOC) than to agriculture and vine growing.

Following the growth and decline of Mont du Cantal, igneous activity backtracked 50 kilometers (30 mi.) to the north and created the Mont Dore volcanic province, starting three million years ago. Activity culminated with the emplacement of the Puy de Sancy volcano, similar in size and eruptive style to Mount Vesuvius, just less than one million years ago. Pyroclastic flows raced downslope and spread volcanic ash in the Allier riverbed — ash that makes up today an excellent terroir around the town of Neschers.

There was a fourth and final bout of volcanic activity in Auvergne, back near the city of Clermont-Ferrand where it all started. A chain of several dozen small cones and domes — similar in size to Sunset Crater in Arizona — popped up in two pulses; the first occurred around thirty to forty thousand years ago, and the second around eight to eleven thousand years ago. Their location high up on the rift shoulder makes for too cool a climate to grow grapes, and besides, this beautiful geological area is protected as a regional park. Most vines are found lower down in the rift valley on more ancient volcanic terrain.

A VOLCANO IN PROVENCE

Southeast of Auvergne, volcanic hills are replaced in Provence by cream-colored limestone laced with rusty beds of clay. The area has long been a stronghold of olive oil and wine production, including the regional appellations Côtes-de-Provence and Coteaux-d'Aix-en-Provence.

There is an interesting geological oddity that stands out in the midst of the limestone hills. Halfway between Aix-en-Provence and the quaint village of Rognes lies the volcanic enclave of Château Beaulieu. It owes its existence to a remarkable episode of plate tectonics that began eighteen million years ago: the rifting and separation of Sardinia and Corsica from Provence, which then swung counterclockwise to their present location. The volcanic rift responsible for the motion has long since disappeared, foundering to the bottom of the Mediterranean, except for the Beaulieu volcano.

Spared by erosion, the crater now stands out as a high-perched plateau, although its early eruptions took place under shallow water—explosive eruptions that created a wide, funnel-shaped crater, approximately 1 kilometer (0.6 mi.) wide, which then filled with lava to a depth of over 100 meters (330 ft.). Later, the crater harbored a lake in which limestone deposits trapped fossils of ancient species, such as a hornless rhinoceros and a pig-sized hippopotamus.

Located inside the crater, Château Beaulieu is the largest estate of Coteaux-d'Aix-en-Provence, with 140 hectares (350 acres) devoted to a standard rosé wine (Château Beaulieu), and 8 hectares (20 acres) reserved for upper-tier wines, going by the different label Villa Baulieu (spelled without an *e*). An elegant sixteenth-century château serves as the estate's headquarters and as a very classy guesthouse. The estate also cultivates almonds and truffles. Wine tasting is conducted in the adjacent winery, and the estate manager Bérengère Guénant organizes occasional hikes around the crater ("wine safaris").

Basalt is conspicuous throughout the crater and crops out mostly on its periphery. The high altitude of 400 meters (1,320 ft.) constitutes a cooler climate than elsewhere in Provence. The larger swing in day-to-night temperatures brings more acidity to the wines, in particular those made from Cabernet Sauvignon (red), and Rolle and Sémillon (white), which are grown on the outer, northern slope of the crater to enhance the cooling effect. Grenache and Syrah, though, are grown in the crater flats, closer to the château. The estate proposes a range of wines that fit all budgets: approximately €10 for a standard red, white, or rosé (Château Beaulieu Cuvée Alexandre); €18 for the upper-tier Villa Baulieu Cuvée Bérengère red or white; and €30 for the red or white Villa Baulieu *grand vin*. The wine house is open for tasting Monday through Friday (8:30 a.m.– 12:30 p.m. and 2 p.m.–5:30 p.m., no charge). (Château Beaulieu and Villa Baulieu, route de Beaulieu, Rognes-en-Provence. Tel.: (33) 4 42 50 20 19; www.chateaubeaulieu.fr/en and www .villabaulieu.com.)

Vines Rooted in History

Auvergne occupies a special place in the history of France. It is here, in the basalt-topped hills, that the Gauls — notably the Arverni tribes led by Vercingetorix — took up arms against Caesar's army in 52 BC.

During this mighty clash, the Romans were initially routed after an unsuccessful siege of Vercingetorix's army, well entrenched on Gergovia hill, south of Clermont-Ferrand. Caesar gained a decisive victory three months later at Alesia, forcing Vercingetorix to surrender (he was later executed in his prison cell) and putting an end to the feud between the Gauls and the Roman Empire.

Historians point out that one of the major motives of the Gallic Wars, apart from the Roman Empire's retaliation for the murder of Roman merchants, was Rome's interest in conquering and securing France's major trade routes that reached northern Europe, with one particular export in mind: Roman wine. Once brought into submission, Gallic tribes were forbidden from planting their own vines as Rome reinforced its monopoly over the lucrative wine market.

The market had been established in France during the second century BC by pioneering Roman merchants, and their work had not been easy: the locals were notoriously fond of *cervoise* — an herb-spiced, barley-based ancestor of beer — and were slow to take up wine. When they finally did, they turned out to be excellent clients, though embarrassingly uncouth: instead of diluting wine with water, as would any proper Roman, the Gauls swallowed it undiluted, which earned them a pretty bad reputation.

Archeological digs support this global view of the wine trade in Gallo-Roman days: vineyards are absent from Auvergne until the first century AD, whereas Roman wine amphorae are conspicuous, notably in the Gallic fortified village of Corent, 20 kilometers (12 mi.) south of Clermont-Ferrand.

The hill of Corent is home today to one of the five "Village" appellations of Auvergne wine: Côtes-d'Auvergne-Corent AOC.[2] It was the access, though, not to wine but to artesian spring water, bubbling up at the summit under pressure, that made the hill a stronghold of Gallic culture. Established as a fortified settlement in the first century BC, it probably was the capital of the Arverni tribes and the birthplace of Gallic war chief Vercingetorix.

Even before the Gallic wars, one can trace the mounting influence of Roman commerce at Corent. Stone constructions were beginning to replace wooden buildings. A large number of wine amphorae with their necks chopped off show up in Corent's sanctuary: wine was probably poured into libation vats as offerings to the gods.

Archeologists also identified several wine shops on the summit plateau, including a tavern and a wine cellar dug into the basalt bedrock. Twenty unbroken amphorae were cataloged, as were thousands of pottery shards, allowing researchers to estimate that Corent's cellar contained at least one hundred amphorae, equivalent to three thousand wine bottles today.

As stated earlier, this was Roman wine, shipped from Rome and its colonies. It was only much later that vines were planted locally and Auvergne broke away from Rome's monopoly. Such local wine is first mentioned in a manuscript written by Sidoine Apollinaire (AD 430–489), bishop of Clermont-Ferrand: it is described as being as good as "the finest Italian wines."

The following few centuries saw the collapse of the Roman Empire and barbarian invasions from the northeast, until Frankish emperor Charlemagne (742–814) unified the country, took over the wine trade, and turned it over to monasteries. There were an estimated 10,000 hectares (25,000 acres) of vineyards in Auvergne by the late 900s.

A good proportion of Auvergne wine was shipped to Paris, despite a complicated itinerary by boat down the Allier and Loire Rivers, when floods made them deep enough for navigation; then across the hills by cart to reach the Loing River; and a second boat ride to reach the Seine, with a last stretch downstream to the capital.

The reputation and value of Auvergne wine offset the shipping cost, and many a French king ordered it to court, including Charles VII (of Joan of Arc fame), and a century later, Henry IV. The latter was a pivotal figure of France's Wars of Religion (1562–1598), famous for his political pledge to develop the economy, until every family in France could afford a chicken for its Sunday meal. The king himself fancied coq au vin—chicken cooked in wine—and made sure the wine came from the Auvergne vineyards of Chanturgue, a hill overlooking Clermont-Ferrand. In recognition of this royal consecration, Chanturgue is ranked today as another Village appellation of Côtes-d'Auvergne wine.

In the days of Henry IV, Pinot Noir and Gamay were the leading grape varieties of Auvergne, as they still are today. They originated in the hills of Burgundy and adapted well to Auvergne, 100 kilometers (62 mi.) to the southwest, which happens to enjoy a similar climate. Gamay, which is easier to cultivate and has a better yield, provides quantity, whereas Pinot Noir is more fragile and finicky but contributes quality.

The commerce of Auvergne wine truly took off when a canal was built at Briare in 1642, linking the Loire and the Seine river systems, and opening up an uninterrupted fluvial shipping route to Paris. In the centuries that followed, this important trade route became highly organized. Logged wood and coal were also in great demand, so that flat barges, named *sapinières*, were built out of pine trunks and loaded with coal and wine. Once in Paris, the cargo was sold and the barges were dismantled to sell the pinewood, eliminating the need to return the empty transporters upstream. This lucrative market was cornered by Auvergne merchants who established trade posts in Paris and became the major providers of coal, firewood, and wine to the capital.

With the inauguration of a railroad line in 1856 connecting Clermont-Ferrand to Paris, shipping wine to the capital became easier yet. Hence, the late nineteenth century was a golden age for Auvergne, especially because other wine regions became afflicted by the phylloxera blight. Landlocked as it was, Auvergne was temporarily unaffected by the grape disease and took advantage of its neighbors' misfortune by increasing its own wine output. In the department of Puy-de-Dôme, the vineyards expanded to 45,000 hectares (100,000 acres) in the 1890s, providing more wine than the entire region of Burgundy.

Auvergne's golden era was short lived. Central France was struck in turn by the phylloxera blight at the close of the century, just as other wine regions were beginning to recover. To make matters worse, a spell of mildew crippled the remaining vines in 1910. Then came World War I, draining men out of the region, followed a decade later by the Great Depression. The vine-planted area in Auvergne collapsed to 12,000 hectares (30,000 acres) and continued to shrink after World War II, as labor further shifted away from the vineyards, drawn by the burgeoning industries of Clermont-Ferrand.

Vanishing from the national market, Auvergne wines suffered a long

---

GRAPES OF AUVERGNE

Gamay, Pinot Noir, and Chardonnay are the three authorized grape varieties of the Côtes-d'Auvergne AOC that survived ruthless competition over the years with less successful varieties, such as the Rhône-native Syrah that did not adapt well to the bout of climate cooling at the close of the Middle Ages.

Pinot Noir rose to prominence during the Renaissance and adapted so well to the region that it took on the name of Auvernat. The grape is well suited to limestone, which it finds in the rift valley, but does not perform well on volcanic terrain. Although its share of the Auvergne vineyard dropped to 20 percent over the centuries, Pinot Noir is now making a comeback as a varietal wine or as a blend with Gamay.

Gamay displaced Pinot Noir over the years as the leading grape in Auvergne, as it is better suited to granitic and volcanic terrain. Introduced in the region during the eighteenth century, it underwent slight transformations under environmental pressures and careful selection by the vine growers to yield an original cultivar, characterized by late budding, which makes it less prone to frostbite; a looser grape cluster; smaller grapes; and faster ripening than in other regions. Both this local cultivar (Gamay d'Auvergne) and regular Gamay make up 70 percent of the vineyard. Gamay is often the sole grape of Côtes-d'Auvergne red and rosé wines, and in blends it always dominates its Pinot Noir partner.

With respect to white wine, the only grape authorized in the Côtes-d'Auvergne appellation is Chardonnay. It is well adapted to the limestone slopes flanking the rift valley and makes up 10 percent of all planted grapes in the region. Sauvignon ranks second among white grapes in Auvergne but can be marketed only as a generic varietal (Vin de France appellation).

---

eclipse. Only now are they making a comeback, thanks to the emphasis placed on terroir, as the Puy-de-Dôme area is blessed with an original mix of volcanic and sedimentary rock, yielding wines with a distinct character.

The Comeback of Côtes-d'Auvergne

Struck by misfortunes in the early twentieth century, Auvergne wine-makers fought back by setting up a cooperative in 1935 that secured from the state their first official appellation: Vin d'Auvergne. This prevented the regional wine industry from totally collapsing, but even the cooperative foundered after a few years. From its ashes emerged a new institution, La Cave des Coteaux d'Auvergne, which centered its efforts on promoting historical terroirs: Madargue (near the city of Riom); Châteaugay, north of Clermont-Ferrand; Chanturgue, within the city boundary; and Corent, to the south. All pledged to live up to their historical reputation and promote quality over quantity. Farther south along the rift zone, the township of Boudes joined the list in 1977. All five received that year a new Village status, technically labeled AOVDQS, for Appellation d'Origine Vin Délimité de Qualité Supérieure.[3]

Encouraged by this promotion into the middle tier of French wines, a new generation of winemakers took over from their elders, rounding their experience in agricultural schools. Abandoned vineyards were reclaimed and restructured for mechanical upkeep wherever possible, and new grape varieties were tested alongside the ruling Gamay, Pinot Noir, and Chardonnay.

Parallel to this revival of independent wine making, the cooperative underwent yet another mutation and was absorbed by the powerful international cooperative Limagrain. The group invested funds to purchase modern wine-making equipment and produce a line of wines showcased in the cooperative's shop at Saint-Verny, at the foot of the Corent hill.

Efforts on both fronts—by both the independents and the cooperative—ended up paying off: Auvergne wines jumped up one more level and received the coveted AOC status in 2010 (Côtes-d'Auvergne AOC).

Totaling less than 1,000 hectares (2,500 acres), with less than half of that (410 hectares) registered as AOC, the wines of Auvergne are still little known and difficult to find outside their production area. The best way to try out the wines is still to schedule a trip to Auvergne and explore a well-hidden province that not only has wine to offer but also remarkable cheeses, volcanic landscapes, medieval castles, and Gothic cathedrals.

Fire Meets Water: The Châteaugay Terroir

North of Clermont-Ferrand lies one of the best examples of the revival of wine making in the region: Côtes-d'Auvergne-Châteaugay AOC (64 hectares, or 160 acres). The wine's reputation goes back several centuries. Under the reign of the Sun King, Louis XIV, a memoir addressed in 1697 to the Duke of Burgundy places the townships of Châteaugay and nearby Pompignat at the top of the charts "in view of their delicate and most enjoyable wines."[4] Much later, between the two world wars, the first appellation of origin awarded in the region was Coteaux de Châteaugay. The distinction was confirmed in 2011, when the village wine received AOC status, as Côtes-d'Auvergne-Châteaugay.

Framed by the village of Châteaugay to the west, Riom and Cébazat

The salt-and-pepper texture of peperite (left), an explosive mix of fragments of white limestone and dark basalt.

north and south, and the Michelin company's test track to the east, the vineyard is spread over a rolling landscape that intersects a variety of rock layers: limestone, basaltic lava flows, and an original mix of both, named *peperite*.

Recall that twenty-eight million years ago, when the Limagne rift zone began to founder, lakes formed on its dropped floor, teeming with plankton and other shell-building life-forms, showering the lake bottoms with calcareous ooze. Layers of limestone formed, tens of meters thick, until the time when the stretching rift zone allowed magma to reach the surface.

Fire and water do not mix. As the molten basalt penetrated the waterlogged limestone, the bottom sediment was ripped apart by powerful steam explosions. The funnel-shaped vents blasted incandescent clouds of magma mixed with limestone fragments. As the pyroclastic flows settled and cooled, they solidified into a cementlike rock of pale limestone studded with pepperlike grains of dark basalt: an aspect that earned it the name of *peperite*.

It so happens that Gamay, the main cultivar of Côtes-d'Auvergne, often develops an aromatic touch of black pepper when it is grown on crystalline bedrock, as it does in the granitic hills of Beaujolais. The same is true of the peperite terroir of Châteaugay, which yields a fitting, albeit coincidental, correspondence between the visual aspect of the soil and the wine's bouquet: besides the main aroma of red berries, typical of Gamay, it is marked by a note of black pepper.

Châteaugay's terroir is famous for its peperite but also for its complexity. Lava flows cover the peperite and limestone in a coating several dozen meters thick—a dark ledge on which the village and its castle were built. The end result is a multilayered cake, carved open by erosion into a basin stepping down from the lava ledge through the limestone strata, down to the peperite layer. The latter is the heart of the appellation, literally and figuratively, as the basin is heart shaped, carved into two lobes by a streambed.

Stéphane Beaujean is a Châteaugay winemaker who is passionate about geology and terroir. Along with several of his fellow winemakers, he would like to see Grand Cru and Premier Cru subdivisions introduced at Châteaugay, with the Grand Cru label covering the peperite heart of the basin. The Premier Cru zone would concern the limestone slopes of its northern rim, with the best sun exposure, south to southeast.

The Côtes-d'Auvergne-Châteaugay appellation only applies to red wines, with Gamay the main variety and Pinot Noir a complement, but one can also find good rosés and whites under the regional Côtes-d'Auvergne label. The whites especially, based on Chardonnay, are surprisingly good. Chardonnay is well suited to limestone—Burgundy being the prime example—so that the white grape is mostly planted on the limestone slope. Châteaugay's white wine features a Burgundy-like character, unctuous with a bouquet of exotic fruit.

The best red and white wines from Châteaugay can be aged several years, and as chance has it, the area's topography and bedrock are well suited for cellars. Peperite in particular, with its cementlike texture, provides solid walls and ceilings. Cellars were first dug under wineries and then, with the surge of production during the nineteenth century, drilled directly into the hill face. These ancient cellars are best seen in the town of Aubières, their entrance decorated with black and white blocks of basalt and peperite, with a ventilation shaft often sticking out of the hillcrest. There are also good examples of peperite wine cellars at Châteaugay, in the appropriately named rue des Caves.

The Hill of Corent

As mentioned earlier, one of the oldest wine cellars in France, loaded with Roman amphorae, was discovered at the summit of Corent hill, some 20 kilometers (12 mi.) south of Clermont-Ferrand. Hence, it is only fitting that the village received special status — as one of the five crus of Côtes-d'Auvergne — in view of its long tradition of wine trade and wine making.

Although the hill and its surroundings produce red and white wine, Corent is mostly known for its rosé. The blush wine's acidity prevented it from spoiling, so it could be shipped by boat and horse-driven cart to cities like Paris.

The Corent hill is ideally located for commerce, with the Allier River wrapping around its base. Villages developed along the river's course, fitted with small harbors, most notably Martes-de-Veyre, which was already famous in the first century AD for its production of pottery and ceramics — basalt and limestone decayed to precious clay in the river's floodplain. In medieval times and onward, with the rise of wine making, coal extraction, and logging, the Allier River and its harbors rose to prominence. By the nineteenth century, as the Industrial Revolution hit France, the piers of Martes-de-Veyre serviced up to 150 boats at a time — all *sapinière* barges, devoted to coal and wine transportation.

Today, railroads and highways have replaced river transport, but riverside villages and vineyards remain and still have their medieval layout: crops and vines planted on terraces across the hill slopes, and old wineries with vaulted cellars dug into the cliff face, their lava-stone porches opening up onto narrow streets.

Lava is prominent at Corent: a thick ledge of basalt caps the hill. Despite its rounded profile, the landform is not a volcano: like many other similar knobs in the Limagne rift valley, the hill is an outlier left over from the pile of lacustrine sediment deposited tens of millions of years ago, then cut back by erosion. On the Corent site, a lava flow descended from the rift shoulder and covered the fragile limestone layers, three million years ago. It was this hard cap of basalt that locally protected the limestone from erosion and left a hill standing.

In a sense, Corent is more of a limestone terroir than a volcanic one, and indeed from afar it resembles Corton, a limestone hill in Burgundy. The vineyard layout, however, is counterintuitive: the eastern flank —

BOUDES, CHANTURGUE, AND MADARGUE

Three out of the five village appellations of Côtes-d'Auvergne are rooted in volcanic soil: Château-gay, Chanturgue, and Corent. The other two, Madargue and Boudes, have a sedimentary substrate.

The northernmost appellation, Madargue, is set on sandy and loamy soil. Silica and clay most likely play a role in the wine's distinct properties, namely a deep color, a floral bouquet—partly due to some Pinot Noir mixed in with the Gamay—and a nice balance overall, with more subdued fruit and spice than in the volcanic Corent and Châteaugay appellations discussed in the main text.

The small Chanturgue appellation (6 hectares; 15 acres) strides the border between Clermont-Ferrand and Cébazat. The slopes are made up of limestone, with a sprinkle of volcanic ash, and face south: the resulting Gamay wine has a bouquet of citrus fruit and aromatic herbs, with a bit of smokiness, and the potential to age several years in the cellar.

Boudes is the southernmost appellation, slightly off the rift axis, up a small dale (D48 road, Couzilloux valley). The overhanging plateau is capped by a ledge of basalt, but the vine-bearing terraces below are composed of oxidized, red calcareous clay. Blending Gamay and Pinot Noir, the wine has strong tannin that betrays a warmer climate—the transition to southern France—with a bouquet of dark berries, the peppery signature of Gamay, and the gamey and smoky notes of Pinot Noir raised on limestone.

normally the best orientation for vine growing—has few vines because it has steep, less manageable slopes. Instead, the village developed along the precipitous hangout, under the lava ledge. Vines were pushed back by the whims of topography toward the northern and southern end of the hill.

In the elevation bracket most favorable for vine growing, ranging here from 400 to 500 meters (1320 to 1650 ft.), about the middle of the hill's slope, the soil is essentially a calcareous clay with a sprinkling of basaltic rubble that crumbled and rolled downslope from the lava ledge. The basalt provides a superficial, dark stony texture to the terroir, especially on the northern side of the vineyard, where the soil is known as *terres noires* ("black earth").

Only rosé wine is entitled to the Corent Village appellation, in recognition of its long history. Corent rosé has a light golden-orange color due to a very brief interaction between the pressed juice and the pigment-containing grape skins. The resulting wine is fairly acidic, which translates to a crisp sensation of freshness. This is particularly true for the north-facing vineyards, whereas the south-facing ones get more sunshine and so produce more sugar and yield fuller, more unctuous wines.

Slope and orientation, bedrock, soil and the microclimate generated by the Allier River thus shape the personality of Côtes-d'Auvergne wine on a local scale. It is also worthy to note how the topography of the rift valley and its uplifted shoulder affects vine growing on a larger, regional scale. The humid, western trade winds are forced to rise along the rift shoulder over its crest of volcanoes and rain out their water along the way: the air that flows down into the rift valley is thus relatively dry—a lack of clouds that results in 1,800 hours of sunshine a year—and warms up on the way down (this is known as adiabatic heating, also called foehn warming). Over the centuries, this favorable climate has encouraged the culture not only of grapevines but of fruit trees as well, while wheat and other cereals are grown in the flatter stretches of the valley floor.

Wine and Pumice: The Neschers Terroir

Farther south along the rift valley lie more basalt-capped limestone buttes—the mix of sedimentary and volcanic soil typical of Auvergne—with scattered vines on their slopes.

One particular terroir is solely volcanic, however, and somewhat unusual. One reaches it by leaving the main road at Coudes village and heading up the rift shoulder, westward, along the Couze Chambon—a stream that springs forth from the base of the Puy de Sancy volcano, following a gully that used to funnel fiery pyroclastic flows, six hundred thousand years ago.[5] At Neschers, the ancient pile of ash, quarried north of the village, is 6 meters (20 ft.) thick. Its total volume probably reached 2 cubic kilometers (0.5 cu. mi.), on the same scale as the ash flow erupted by Mount St. Helens, in Washington State, in May 1980.

In the early 1900s, vines covered a significant fraction of the valley slopes and plateau, north of Neschers. Today, vine lots are much more scattered, but they yield classy wines, influenced by their pyroclastic, fine-grained, and silica-rich terroir. Toasted and roasted aromas stand out, recalling, mostly coincidentally, the volcanic, "burnt" nature of the land. At Neschers's Miolanne estate, winemakers Jean-Baptiste Deroche and Laure Cartier also point out that their Pinot Noir wine has a floral, rose bouquet on pumice soil, whereas the varietal boasts a more classic fruity and jammy bouquet on their limestone lots.

To complement this general presentation of Côtes-d'Auvergne wines, the itineraries below provide a more detailed look at their setting and characteristics.

Guide Section: Visiting Auvergne

Geological Itinerary

Auvergne is an open-air museum of volcanology. The Cantal extinct volcano, for starters, deserves its own hiking trip, as do Puy de Sancy and the other Monts Dore volcanoes. The introductory tour below focuses on the more recent Chaîne des Puys volcanoes that line the rift shoulder, above the Limagne valley and its Côtes-d'Auvergne vineyards.

There are close to fifty cones and domes built up along the major fault zone, which trends north-south, with most eruptions staggered between fifty thousand and eight thousand years ago. Two of the more recent cones are Puy de la Vache and Puy de Lassolas, which erupted around 7,000 or 6,000 BC. They rise approximately 150 meters (450 ft.) over the plateau and are made of black and red scoria, piled up by lava fountaining and Strombolian eruptions. Lava flowed out to the southeast, keeping that sector of the cones unbuilt and acting like a conveyor belt, carting away the scoria raining down from the sky. Hence the cones are breeched and display a crescent shape.

Other landmark volcanoes are accessible along road D941B ("Volcano Row"), which crosses the chain between the villages of Orcines and Pontgibaud: Puy de Pariou, Puy de Côme, Grand Sarcoui, and Puy Chopine — all are reachable by hiking trails.

An interesting pair to visit is Puy de Pariou and Grand Sarcoui: a cone and a dome. There is a parking lot (Parking des Goules) opposite the road from Puy de Pariou. Cross the road, then a meadow, along a path marked by yellow on trees and stones, and ascend the northern flank of the cone. The base of the volcano was once occupied by a lava lake that emptied out toward the road, accounting for the cone's slightly breeched profile. The 200-meter (660 ft.) cone bears a central crater, 100 meters (330 ft.) deep. From the summit, there is a good view of other volcanoes nearby, including Puy de Côme to the west and Puy de Dôme — the highest peak of the chain — to the south.

After hiking down the southern slope, return to the road by circling the base of the volcano (clockwise or counterclockwise). This first hike takes approximately two to three hours. Cross the road to the parking lot for the second hike (also two hours), following the trail that leads to Grand Sarcoui. The latter is a 220-meter tall (725 ft.) dome, slightly flattened at the top, that grew out of the ground like a giant mushroom. It is a textbook example of a cumulodome: an extrusion of viscous magma — trachyte in this case — that grew in place without flowing outward. The growth was accompanied by flank failures and pyroclastic avalanches, especially on the eastern side.

The trail climbs up the southern flank, leading to a grotto-like, medieval quarry that was dug into the volcano to extract its clear-colored, granular trachyte. The rock was used in Merovingian times (sixth and seventh centuries) to produce stone coffins, and one unfinished sarcophagus can be spotted, half extracted, in the grotto's wall.

Off of Volcano Row (road D941B), there are two educational sites worth visiting: Vulcania and Volcan de Lemptégy. Vulcania is a volcano theme park, built underground, that comprises several exhibits, a make-believe magma chimney and lava pond, as well as an IMAX theater and a 4D Dynamic Cinema.[6] A few hundred meters down the road, Volcan de Lemptégy is less artificial. The park is the remnant of a cinder cone quarried down to almost ground level. It shows in cross-section layers of ash, fallout "bombs," and old fumarole fissures. The owners set up an educative trail around the quarry, with a little train, a 4D Dynamic Cinema, and other attractions.[7]

Puy de Dôme, another cumulodome of trachyte and the tallest volcano of the chain (it's eleven thousand years old), is also worth a visit. The summit, serviced by a rack railway,[8] harbors a TV relay station, a weather station, and Gallo-Roman ruins of a temple dedicated to Mercury. It offers a spectacular view of the chain of volcanoes, especially to the north, with the nearby cinder cones of Puy de Côme and Puy de Pariou. To the east, the view is down the rift shoulder over the Limagne valley and the city of Clermont-Ferrand. The summit is also a popular takeoff spot for hang gliders, who run off the steep slope to soar over the cinder cones below.

In the nearby village of Volvic, light-gray trachyandesite was mined for the construction of many local monuments, including Clermont-Ferrand's cathedral. One can visit the underground quarry and its sound and light show on the history of the stone industry.[9]

Wine Tour North: Volvic, Châteaugay, and Gergovie

The city of Clermont-Ferrand is the hub of the first set of vineyards to visit in Auvergne: Volvic on the slopes of the rift shoulder, Châteaugay to the north, and Gergovie to the south.

Volvic is mostly known for its building stone (*pierre de Volvic*) and its mineral water, but it used to be one of the many wine-producing villages in the area. Today there is only one estate left, revamped by organic winemaker Vincent Marie.[10] His love of nature is obvious in the way he describes his five lots on different soil and bedrock (granite, basalt, limestone), and his environmentally friendly tilling techniques (for one lot, he uses a horse). Vincent Marie produces organic "natural' wines, with no commercial yeasts and no sulfites or preservatives, so the result can be quite surprising. He focuses mainly on Gamay (some vines are over one hundred years old), Pinot Noir, and Chardonnay, with wine names that are puns reflecting his love of music, including Rockaille Billy, Fusion, and Magma Rock. Gamay and Pinot Noir are mostly grown on granitic sand and basaltic scoria, which gives the red and sparkling rosé wines the spicy, smoky notes typical of volcanic wine in Auvergne. The rift shoulder towers over one of Vincent Marie's lots, crowned by the medieval castle of Tour-noël. Visits upon request.

Châteaugay, close to Clermont-Ferrand, is another village that boasts a medieval castle. Built in the fourteenth century, the castle now hosts a restaurant and a wine shop that goes by the friendly name of Caveau du Gay Coeur ("Cellar of the Joyful Heart"), dedicated to the Village appellation: Côtes-d'Auvergne-Châteaugay AOC. The vineyards are draped over a rolling landscape of limestone, basalt, and peperite tuff; some are planted on walled terraces, interspersed with fruit trees.

About a dozen estates share the appellation (100 hectares, or 250 acres) in Châteaugay and the nearby villages of Blanzat, Pompignat, Cébazat, Ménétrol, and Riom. My favorite winemaker, based in Blanzat, is Stéphane Bonjean, last in line of a family of winemakers and passionate about geology. Stéphane watches over 6 hectares (15 acres) of vines, distributed over the many different terroirs of the appellation, and tends half a dozen wine cellars carved into the basalt and peperite bedrock. His wines include Côtes-d'Auvergne and Côtes-d'Auvergne-Châteaugay (mostly reds), including special vintages named after his children: Cuvée Élisa and Cuvée Gabin.[11]

South of Clermont-Ferrand, the hill of Gergovie hosts another interesting terroir that blends geology, history, and wine making. To reach it from Châteaugay, one can drive through Clermont-Ferrand and visit the city-enclaved Chanturgue vineyard along the way or take the highway and bypass the city (exit at La Roche–Blanche).

The Gergovie outlier is a textbook example of inverted topography: the site was once a valley but now lies above the surrounding terrain. First, lacustrine sediment and volcanic tuff (peperite) were deposited in the ancient valley. Next, basaltic lava flows covered up the valley floor nineteen million years ago, protecting it from erosion. While streams were later able to wear down the unprotected valley flanks, the basalt-covered floor was spared, eventually ending up higher than its worn-down surroundings.

On account of its steep slopes, the basalt-covered Gergovie plateau was among the major strongholds of the Gallic civilization. During the Gallic Wars, tribal leader Vercingetorix entrenched his army in the fortified city on the summit (then known as Gergovia) and routed Cesar's army by conducting a number of damaging infantry and cavalry charges.

Today, another rebel stands up to imperialist forces. Fighting the region's powerful agricultural holding that controls most of Auvergne's wine market, Gilles Persilier vociferously defends small-scale organic wine making and refuses to use industrial pesticides or nonlocal yeasts.[12] From his winery on the hillcrest, he turns out a wide range of red, white, rosé, and sparkling wines, with names that evoke the glorious days of the Gallic tribes, such as Cuvée Vercingetorix, in honor of the charismatic leader (Gamay d'Auvergne and Pinot Noir), and Cuvée Celtil, named after Vercingetorix's father (Gamay d'Auvergne Vieilles Vignes).[13]

Wine Tour South: Corent and Neschers

About 10 kilometers (6 mi.) south of Clermont-Ferrand stands the hill of Corent, another pile of lacustrine sediments capped by a lava flow that protected the site from erosion. It too was crowned by a Gallic fort (see pages 127–28), presumably the birthplace of Vercingetorix. Archeological digs are conducted on the summit during the summer. Visitors can tour the ruins, including shattered Gallo-Roman wine amphorae in the central forum.

Below the summit, the present-day village of Corent clings to the lava ledge, its houses and cellars dug directly into the basalt bedrock. Most producers of Corent rosé wine have now set up shop down in the valley, but visitors should check out the cellar of Christian Liévain (by appointment),[14] where the winemaker stores antique tools, an intact Gallo-Roman amphora, and a well at the back of the cellar where water gushes up from an artesian aquifer — a geological asset that made Corent a siege-resistant stronghold in the past.

Besides the independent winemakers established in the valley (Marc and Jean-Pierre Pradier at Martes-de-Veyre, and Thierry Sciortino at Vic-le-Comte, to name but a few), the Côtes-d'Auvergne wine cooperative established its headquarters and showcase wine shop at the foot of the Corent hill: Cave Saint-Verny, on the outskirts of Veyre-Monton.[15]

There is one more village worth visiting for wine, farther south along the Allier River. Follow country roads or the A75 highway (pulling off at exit 8) to the village of Coudes, then climb west along the Couze Chambon tributary (road D229), up the rift shoulder to the town of Neschers.

The picturesque village is built on a basalt flow at the foot of a steep slope where vines once stood. Driving through the village, over a bridge and up the slope, one crosses a layer of light-colored pumice near the top of the plateau. A roadside quarry shows a cross-section of the ash layer, some 6 meters (20 ft.) thick, that rolled down as a hot pyroclastic flow from the Puy de Sancy volcano on the horizon, five hundred thousand years ago.

It is here that winemakers Jean-Baptiste Deroche and Laure Cartier recently took over a successful vineyard, Domaine Miolanne, first sharing the experience and winery of the previous owners, then building their own ecologically oriented, larch-wood winery along the road,[16] closer to their 6 hectares (15 acres) of vines on the plateau, set both on clay-limestone soil and on the silica-rich, volcanic ash flow. Wines from the latter are the finest: a quality provided by the soil's texture and chemical makeup, as well as by the wind blowing down from the Puy de Sancy ridge in the distance, which cools the grapes and lengthens their ripening period.

To finish the day, one can drive upstream toward the volcano and through the village of Saint-Nectaire, famous for its cheese; or return to the Allier main valley and visit the next, parallel tributary up the rift shoulder, Couze Pavin, to visit the troglodyte dwellings dug into the ash flows outside the hamlets of Perrier and Saint-Pierre-Colamine (Grottes de Jonas).

CHAPTER 7

The Canary Islands

The Spanish Canary Islands, off the coast of Morocco, show another original aspect of the interplay of volcanoes and wine. In the island of Lanzarote, overrun by a spectacular eruption in the eighteenth century, vines are planted in funnel-shaped depressions, dug into the basaltic rubble. Based on Malvasia, Listán, and Moscatel grapes, Canary wines encompass sweet and dry whites as well as fruity, peppery reds.

The Canary archipelago forms a crescent of islands stretching from Lanzarote and Fuerteventura in the east, 100 kilometers (60 mi.) off the coast of Africa, through Gran Canaria, Tenerife, and La Gomera, and out to La Palma and El Hierro to the west, 500 kilometers (300 mi.) from the mainland. The genesis of this volcanic chain is not fully understood, but it is probably linked to a hot spot in the underlying mantle, through deep crustal faults that help the magma reach the surface.

The Canary Islands were built up over extensive periods of time, in contrast to the Aeolian Islands, Mount Vesuvius, or Mount Etna, all of which are younger than five hundred thousand years old. The oldest lavas in the Canary Islands reach thirty-five million years in Fuerteventura, twenty million years in Lanzarote, and fifteen million years in Gran Canaria, Tenerife, and La Gomera. Only El Hierro and La Palma, at the western end of the chain, are substantially younger, emerging less than a couple of million years ago.

Volcanic activity is ongoing today. Although no eruption has taken place in Gran Canaria, La Gomera, or Fuerteventura in historical times

Satellite image of Lanzarote, the easternmost island of the Canaries.

(since Spanish colonies were established at the beginning of the fifteenth century), the most recent lava flows are no older than a few centuries. All other islands were rocked by historical eruptions: El Hierro experienced one in 1793; and Lanzarote, Tenerife, and La Palma experienced no fewer than seventeen eruptions, the last occurring in Lanzarote in 1824, Tenerife in 1909, and La Palma in 1971. It is obvious from this list that the next eruption in the Canary Islands should occur in the near future, as the interval between eruptions is forty years on average.

Eruption styles are quite varied in the Canary Islands, in step with the chemical composition of the magma, which ranges from a fluid, silica-poor melt (basalt) to a silica-rich, viscous, and explosive paste (trachyte and rhyolite). Basaltic magma erupts as continuous Hawaiian lava fountains or intermittent Strombolian lava spurts, building up ash cones and flooding large areas with streams of fluid lava, whereas the less frequent, pasty eruptions build up domes with short, stubby flows of rhyolite and obsidian, or occasionally blow out gas-rich clouds of hot pumice.

One explanation for this varied behavior is that some volcanoes funnel magma directly to the surface, yielding basalt, whereas others possess a magma chamber below the surface, where the rising melt stalls, cools, and decants, dropping metal-rich crystals out of solution while a foam of silica and gas builds up at the top of the reservoir to fuel the more explosive eruptions.

Vineyard History and Distribution

Settled during the fifteenth century by Spanish and Portuguese navigators,[1] ideally located on the trade-wind route between the Old and New Worlds, the Canary Islands served both as an agricultural colony for Iberia and as a waypoint to stock up food and wine for ships bound for the Americas.

Grapevines and sugarcane were the first food stock cultivated, but the sugar industry lasted little more than a century, phased out by the rival colonies of Brazil and the Caribbean Islands.

The Canary wine industry fared much better. England became a faithful client for both dry and sweet white wine, produced from the same Malvasia grape cultivated in the Aeolian Islands off the coast of Sicily. The wine was praised by English doctors for its medicinal value, notably to fight tuberculosis and diabetes. Shakespeare had several of his characters mention it, including the innkeeper Mistress Quickly, who scolds one of her regulars in *Henry IV*: "You have drunk too much canaries; and that's a marvelous searching wine, and it perfumes the blood."

Mostly produced on Tenerife Island at the time, two thirds of the Canaries' wine production was shipped to London by the end of the seventeenth century, but the industry eventually collapsed, as commercial ties between Spain and England were ruptured and British courts turned to Portugal's Madeira Islands for their wine supply. The Canary Islands' wine industry was further weakened by two waves of fungi that struck the vineyards: powdery mildew in 1852 and downy mildew in 1878.

Canary wine never recovered on the international scene, but it did manage to scale down and reach a new balance, catering to the local market and a growing influx of tourists. Today, roughly 20,000 hectares (50,000 acres) of vine are cultivated in the Canary Islands (3 percent of the archipelago's total area). The main island of Tenerife ranks first (12,000 hectares), followed by Lanzarote (3,500 hectares, or 8,600 acres) and La Palma (3,000 hectares, or 7,400 acres). The remaining half dozen islands account for only a few hundred hectares of vines, supplanted by banana, tomato, and potato crops.

Canary wines were granted DO appellations (Denominación de Origen) starting in 1992. The main island of Tenerife includes Tacoronte-Acentejo (the first DO recognized in the Canaries) in the lava fields of its northeastern peninsula; Valle de la Orotava, on basaltic ground at the foot of the Teide volcano; Ycoden-Daute-Isora, extending it to the west; Abona to the south, where vines climb the slopes of Teide volcano, up to an elevation of 1,750 meters (5,250 ft.); and Valle de Güímar to the east.

There are five other appellations in the archipelago, all named after their respective islands: El Hierro, La Palma, La Gomera, Gran Canaria, and Lanzarote. The latter is known for its spectacular landscape, born of a mighty eruption, close to three hundred years ago.

Lanzarote and the 1730 Eruption

Contrary to its neighbors Fuerteventura and Gran Canaria, Lanzarote had not witnessed an eruption since the establishment of Spanish colonies in the 1500s. The truce came to an end on the evening of September 1, 1730, when a long fissure cracked the ground outside of Timanfaya village. Fountains of magma blew skyward, feeding streams of lava that raced downslope toward the sea, "as fast as flowing water," according to Don Andrés Lorenzo Curbelo, priest of the neighboring village of Yaiza, who described the eruption in great detail in his journal.

The abruptness of the eruption and the unusual fluidity of the lava are credited to its origin deep in the mantle, as a hot, metal-rich melt that rose briskly to the surface, with no time to cool and become viscous along the way. It brought up fragments of green rock ripped from the mantle (olivine peridotite), as well as deep-seated gas that fueled the eruption and pulverized a large fraction of the magma into scoria and fine ash.

Throughout the month of September, one fissure after another broke the ground and poured out streams of lava, overrunning farms and villages along the way. After a short pause, the eruption resumed in mid-October. A rain of scoria and ash blocked out the sun and spread its dark shroud over the fields, burying crops and choking cattle in the process.

Bodega (winery) La Geria, with vines at the bottom of a bowl dug into the ash fall from 1730–1736.

The eruption continued all winter long and rolled into the following year, setting more villages ablaze and forcing the population to abandon the island and seek refuge in Gran Canaria. When the eruption finally ceased, in April 1736, 200 square kilometers (80 sq. mi.) of cropland — a quarter of Lanzarote's total area — were buried under several meters of basalt, with at its center a row of thirty-two cinder cones lined up over a distance of 18 kilometers (11 mi.).

This central fissure zone of La Geria rose from its ashes to become two centuries later the greatest and most astonishing viticultural area of the Canary Islands.

Holes in the Ash

Five years after the end of the great eruption, the population returned to Lanzarote, to face a desolate, lunar landscape, with rare sprigs of vegetation poking out of the ash. Farmers got back to work and dug funnel-shaped holes through the scoria to reach the buried soil, planted seeds, and cuttings, and covered them up with a few centimeters of basaltic sand.

This *arenado* technique ("sand burial" in Spanish) proved highly efficient. Not only did the hole reach fertile ground; it also offered protection from the wind and helped trap the little water available.

Because there are no mountains on the island capable of deflecting the humid ocean air high enough for its moisture to condense, Lanzarote is a dry desert, with only 150 millimeters (6 in.) of rainfall a year. But it so happens that the volcanic scoria, named *picón* in the Canary Islands, is ridden with holes: it traps moisture like a sponge, soaking up the morning dew. In Lanzarote, the water stored in this manner equates to an extra 60 millimeters (2.5 in.) of rainfall a year: a 40 percent supplement, with respect to the 150 millimeters (6 in.) that rains down directly from the sky. Moreover, because rainfall is rare but often torrential, the spongelike texture of scoria helps the water penetrate the ground rather than scour the slopes and strip off the little soil present.

Lanzarote is also struck by strong winds: the fine ash carried aloft blasts flowers and young plants, as it also does in Greece's Santorini. Here, vine growers adopted a different solution than the protective ring-shaped pruning conducted on the Greek island. In Lanzarote, not only are vines planted at the bottom of bowl-shaped dugouts; they are also further protected by a rampart of stones arranged around the crest of the bowl, to prevent sand from sliding in and covering up the vines.

The number of these *hoyos* ("holes" in Spanish) has expanded since World War II, with often two or three vines per hole. The result is an aesthetic landscape of evenly spaced holes with stone ramparts. The motif is particularly spectacular next to La Geria's eruptive vents, where the ash cover is thickest and the holes correspondingly larger, to reach the buried soil. Here, the stone rim around each bowl is a crescent facing northeast: the trade-wind direction. Its light-colored lava stone contrasts sharply with the dark scoria and adds an extra touch to the artistic array.

At La Geria, it is not uncommon for the *hoyos* to reach a depth of 2–3 meters (6–9 ft.), and a diameter roughly double that. In other places, where the blanket of ash is less than 20 centimeters (8 in.) thick, vine growers don't bother digging out a bowl: they simply punch a hole through the ash and plant the vine with no further protection.

For medium-sized bowls, spaced 10 meters (30 ft.) apart in staggered rows, one reaches a density of approximately seventy to one hundred *hoyos* per hectare (about twenty-eight to forty *hoyos* per acre). With only two or three vines per hole and three or four grape clusters per vine, yields are naturally limited: 400 liters of wine per hectare, whereas in a standard vine-growing region like Bordeaux or Napa Valley yields are ten to twenty times higher. However, there is plenty of room in Lanzarote's volcanic desert, and no need to maximize yields.

The success of the *arenado* or *hoyo* method has spread to other islands in the Canaries. Although vineyards elsewhere do not have natural fields of deep scoria comparable to those in Lanzarote, in the 1950s winemakers began to create them artificially by quarrying old cinder cones and spreading the scoria on their lots.

The scoria layer is kept thin enough for it to achieve its heating and wind-protection role, without the need to dig a funnel-shaped bowl for each plant. And because they fully control the substrate's makeup, farmers often spread manure on the ground before covering it up with scoria. With the fertilizer thus incorporated, the technique is successful not only for grapevines but also for onions, potatoes, and other popular crops of the Canary Islands.

Grape Varieties in the Canaries

There are about fifty grape varieties in the Canary Islands, a surprising number in view of the archipelago's small area of 7,500 square kilometers (3,000 sq. mi.), about the size of Delaware. About twenty of them are indigenous and found nowhere else in the world. Such diversity stems from a long history of colonization and the introduction of many cultivars since the fifteenth century. The dry climate and the soil's volcanic texture also protected the grapes from disease. The phylloxera louse, in particular, never managed to gain a foothold in the Canaries, perhaps because the sandy, dry volcanic soil is unfit for the insect, as is also the case in Greece's Santorini. For this reason, most Canary varieties are ungrafted—no need for phylloxera-resistant American vine stock—and can reach a venerable age.

The half dozen varieties that account for most of the wine produced in the Canaries originally come from Spain and elsewhere around the Mediterranean basin. They are both red and white grapes, and their proportion varies from one island to another, with the total production in the archipelago averaging 53 percent red wine, 43 percent white, and 4 percent rosé.

In Lanzarote, the emphasis is on white wine, with Malvasia the dominant variety, and Muscat and Listán raising the total share of white grapes to 75 percent. Malvasia is also a popular grape in the southwestern section

THE AZORES ISLANDS

Much farther out in the Atlantic Ocean than the Canary Islands, the Azores Islands are close to the mid-ocean ridge, lined up along a major east-west fault zone (Azores-Gibraltar) that splits the Atlantic basin into two tectonic plates (Eurasia and Africa): a setting known as a "triple junction." The intersection of these fault zones explains the extraordinary volcanism that raised the Azores out of the waters: an archipelago of nine islands, with the last two — Faial and Pico — less than a million years old. Since the Azores were settled in the fifteenth century, the historical record shows twenty-eight eruptions: fifteen above water and thirteen submarine, the last one shooting scoria out of the water off the coast of Faial in 1957 (Capelinhos volcano).

The first missionaries who reached the islands recognized the similarities between the geology and climate of the Azores Islands and those of Sicily, and imported Sicilian grape varieties, namely Verdelho. From it, Azorean winemakers produced a sweet white wine that became popular in the eighteenth century and reached the courts of Europe and Russia. Unfortunately, an epidemic of powdery mildew in 1852 (the same one that hit the Canary Islands) devastated the vineyards. Few varieties survived, and new ones were introduced, including the early American strand Isabella, and more recently the popular varieties Merlot and Cabernet Sauvignon.

The vine-planted landscapes of the Azores have their own special layout that earned them recognition as a UNESCO World Heritage Site in 2004: the small lots are separated by dry walls of basalt named *currais* that protect the vines from the wind. The dark ledges also act as solar panels and radiators, speeding up the ripening of the grapes: a welcome contribution for the production of sugar-rich sweet whites.

The island of Pico accounts for most vineyards in the archipelago: it is known for its sweet white Verdelho de Pico, recently renamed Lajido. The island of Terceira also produces sweet whites, and Graciosa specializes in light and fruity dry whites. There also are a few red wines in the Azores made from Merlot and Cabernet Sauvignon, with evocative names like Terras de Lava and Basalta.

of La Palma (region of Fuencaliente–Las Manchas), and a minor player in Gran Canaria and Tenerife, where the grape is challenged by Listán, as well as by easier, more profitable crops such as bananas. Tenerife's Malvasia is similar to the variety grown in the Aeolian Islands, but Lanzarote's version developed original traits through crossbreeding with the local Marmajuelo variety.

White Listán (also known as Palomino Fino) is famous in Spain as the primary grape of Andalusia's sherry (Xeres) fortified wine. It is the number-one variety of Tenerife vineyards. Other white varieties include Gual; Forstera Blanca; and Diego (also called Vijariego), a grape that ranks high both in sugar and acidity, making it a good candidate for oak-barrel aging and also for sparkling wine (*espumoso*).

The leading variety of red grapes is Listán Negro, the dark cousin of white Listán, brought to the Canary Islands by the first settlers.[2] It is the most planted variety today, red and white grapes combined, and covers a total of 5,000 hectares (12,500 acres), yielding a red wine with fruity aromas of black cherry and black currant, and notes of black pepper.

Claiming second place in the red grape category at 1,200 hectares (3,000 acres), Negramoll originated in Andalusia and was first established in Madeira, where it became the leading variety of fortified wines. The success of Madeira wines motivated Canary winemakers in the eighteenth century to plant Negramoll as well, to compete with the Portuguese. The grape has a rather soft skin — hence its name, which means "soft black" in Spanish — and yields a low-tannin wine with a light, Burgundy-like color and a bouquet of cherry and wild strawberry. Its low-key yet aromatic character makes it a good partner in red wine blends.

Finally, among minor red varieties, is Tintilla, which yields a wine with a bouquet of dark berries and chocolate. It is probably related to the Trousseau variety cultivated in the Jura Mountains of eastern France.

Canary Wines

Lanzarote and La Gomera perpetuate the tradition of Canary white wine, whereas Gran Canaria and the two main provinces of Tenerife (Abona and Tacoronte-Acentejo) have opted for red. Tenerife's three other provinces and the islands of La Palma and El Hierro are split between both. As for rosé, there is a small percentage produced across the entire archipelago, and even some sparkling wine made in Tenerife's Güímar Valley and in Lanzarote.

The diversity of Canary wines is amplified by the high number of estates: more than nine thousand vine growers share the 8,000 hectares (20,000 acres) of vines classified as DO, which translates into less than a hectare per owner. Most small owners, however, sell their harvest to the larger estates, so that the wine production is in the hands of 330 wineries, or bodegas. The latter stay modest in size in La Gomera and Gran Canaria at less than 10 hectares (25 acres) per bodega, whereas those in Lanzarote average closer to 100 hectares (250 acres).

Malvasia, the leading white wine, is produced and marketed in several versions: sweet or dry, and raised in steel or raised in oak. The dry version, Malvasia Seco, exhales an herbal and fruity bouquet, and has an unctuous texture with a warm feeling to it; the sweet version, Malvasia Dolce, is more complex, with a bouquet of dry herbs, mint, and white flowers, and on the palate more amplitude and a fresher bite, due to a better balance of acidity and residual sugar. The oak-raised version (Malvasia Seco *fermentado en barricas*) is closer to Madeira wine, with woody, toasted aromas.

White wine made from Listán Bianco (often assembled with auxiliary varieties such as Albillo, Moscatel, or Malvasia), has a dry, acid character, with a neutrality that brings out the minerality of the volcanic terroir and lets other varieties in the blend express their subtle aromas.

Finally, there is a small production of fortified wines in the Madeira style made with Muscat of Alexandria (Moscatel de Alejandría). This *vino de licor muscatel* boasts aromas of hazelnut, honey, and fig, and reaches 15 to 22 percent alcohol.

Rosés (*rosados*) constitute a minority (5 percent) of Canary wines. Made from Listán Negro, they take on a variety of shades, from orange to salmon pink, with strawberry as their main bouquet.

Red wines (*tintos*) are based on Listán Negro, Negramoll, or a combination of both, with occasionally a splash of white Listán. They can be raised in steel vats for quick consumption (*tinto tradicional joven*), raised in oak (*tinto tradicional pasado por barrica*), or vinified Beaujolais-style by carbonic maceration of unpressed grape clusters (*tinto maceración carbónica*).

The more traditional vinification techniques yield cherry-colored wine with aromas of black and red berries, pepper and other spices. Oak adds jammy and toasted notes. Carbonic maceration shifts the wine color to purple and the bouquet to strawberry, with a more velvety, rounded palate.

According to oenologists, many Canary wines include a mineral component that betrays their volcanic terroir and a pinch of iodine and salt that highlights the vineyards' proximity to the sea.

Guide Section: Visiting the Canary Islands

The Canary Islands are reached by plane from Spain (two and a half hours from Madrid) and from a number of European cities. Hydrofoils are a convenient way to hop from island to island (Tenerife to La Gomera takes forty minutes, for instance), although interisland flights are more practical for the longer hauls, such as Tenerife to Lanzarote (ferries take ten hours).

The oceanic, subtropical climate is pleasant all year round, cooled by trade-wind northerlies. The grape harvest begins in July, and there is a wine fest on August 15 in Lanzarote. Each island has its own charm: fine sand beaches and good scuba diving (with a chance of running across dolphins and sea turtles) in Fuerteventura and primordial forests in La Gomera (Garajonay National Park). But for wine lovers, the islands to visit are Tenerife and Lanzarote.

Tenerife

Tenerife possesses the most vineyards — spanning five subarea appellations — as well as the tallest volcano in the archipelago, Mount Teide, which last erupted in 1909. The volcano rises 3,718 meters (12,280 ft.)

above sea level, and given its large underwater pedestal, it ranks as the third tallest volcano on the planet, behind Hawaii's Mauna Loa and Mauna Kea, as it peaks 7,400 meters (24,600 ft.) above the ocean floor. It also holds another record: the highest vineyards in Europe, which climb up to 1,700 meters (5,800 ft.).

There are approximately two hundred bodegas on Tenerife, which is enough to keep wine amateurs busy, but for an overview, a good address is the Casa del Vino La Baranda,[3] north of the volcano. The old winery has been turned into a wine museum, restaurant, and boutique. The terrace offers a great view, with the ocean to the north and Teide volcano to the south, and a vine garden displays the major cultivars of the Canary Islands. There are more than three hundred different wines sold on site, and about two dozen are featured each day for wine tasting, at an average cost of €2 a glass. Bottles on sale are in the range of €10 to €20.

Lanzarote

Anyone interested in volcanoes and wine should visit Lanzarote and the rift zone of the 1730–1736 eruption. The black and red lunar landscape is studded with over thirty cinder cones, now part of the Timanfaya National Park. Access is restricted to preserve the site, and no hiking is allowed within the park's boundaries. Tour buses depart from the visitor center's parking lot; the tour lasts forty-five minutes, with windows closed and no getting out of the bus. The view, nonetheless, is spectacular.

Outside the park, the main road that runs along the southern edge of the lava field, northeast to southwest, offers a spectacular view of older cinder cones and of the vines in their funnel-shaped pits of black scoria. Here, one can walk up a number of paths along the flanks of the cones to take good photos of the vineyards.

All along LZ-30, as the road is known, there are a dozen bodegas to stop and taste some of the finest wines in the Canary Islands. To reach the "strip," drive out of Arrecife—Lanzarote's main city and airport—toward the northwest on LZ-20, via San Bartolomé, for approximately 10 kilometers (6 mi.), to the intersection with LZ-30. Take a left on the latter, southwest across the lava fields. About 2 kilometers (1.5 mi.) down the road, after crossing the hamlet of La Florida, a small road branches off to the right, along the white wall of a private property (La Alborada), and leads to Los Bermejos estate 800 meters (2,600 ft.) down the road. A modern venture relying on traditional wine-making techniques, Los Bermejos turns out a dozen excellent wines, including four Malvasia and a Moscatel, a dry Diego, two Listán rosés (including one sparkling version) and two Listán reds.

Back on the main wine road, LZ-30, one can now admire the rows of *hoyos* and their stone walls for a few kilometers, before reaching on the right of the road Bodega El Grifo,[4] one of the island's oldest estates. Its wine cellar carved into the lava flow has been transformed into a wine museum, with guided tours at fixed hours (once or twice a day; check the schedule ahead of time) of the museum, the modern winery, and the vineyard, and including the tasting of two wines, all for approximately €15. It was El Grifo that introduced most modern wine-making techniques to the Canary Islands, including cold pressing (cryoextraction) and stainless-steel vats.

Another long stretch of 8 kilometers (5 mi.) along LZ-30 leads to La Geria estate. Just before reaching it, a quaint option for lunch is the small Bodega El Chupadero,[5] 200 meters (650 ft.) up a small road to the left. The restaurant is open every day, except Monday, from 11 a.m. to 11 p.m. (the kitchen closes at 9 p.m.), and serves local dishes and its own house wine (credit cards are not accepted). After lunch, drive back down to the main road and take a left to La Geria, a few hundred meters ahead. Notice the chapel of Nuestra Señora de la Caridad: the church, founded in 1699, was buried under the scoria of the 1730–1736 eruption but later cleared of its ash cover. Competing for your attention, Bodega La Geria,[6] and Bodegas Rubicón across the street,[7] will entice you to return to the pagan world of wine tasting.

CHAPTER 8

California, Oregon, and Hawaii

Two of the greatest wine terroirs on the planet are found in the western United States: Napa Valley in California and Willamette Valley in Oregon. Underlying their celebrated estates, adjacent volcanic and sedimentary rock make it possible to investigate the contribution of their contrasting soil types to the quality and personality of the regions' wines.

The US Pacific Northwest is a showcase of volcanism and plate tectonics, boasting both the San Andreas Fault in Southern California, where the Pacific and North American plates slide past each other, and the Cascade Range of volcanoes to the north, where the oceanic plate dives under its continental neighbor, fueling explosive eruptions — such as the cataclysmic awakening of Mount St. Helens in 1980.

Eruptions on an even grander scale took place in the Pacific Northwest some seventeen million to fourteen million years ago, covering wide areas of what is today Idaho, Oregon, and Washington under thick lava flows. This type of rare but monumental volcanism, known as continental flood basalts or traps,[1] occurs every few tens of millions of years. The western US occurrence, known as the Columbia River Basalts, is the last on record, the previous ones being the Ethiopian traps (thirty million years ago), North Atlantic traps (sixty million years ago), and the Deccan traps of India (sixty-seven million years ago).

The Columbia River lava fields were so voluminous, hot, and fluid that they spread like motor oil, building no feature as familiar and spectacular as the volcanoes described so far in this book but underpinning some

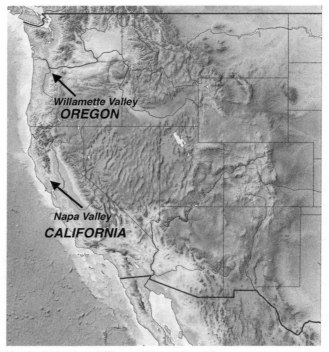

Map showing locations of Willamette Valley and Napa Valley.

of the best wine estates in Oregon's Willamette Valley. Although the lava fields did not reach California, similar-aged basalts also erupted in Napa Valley, one of America's finest volcanic terroirs.

North of San Francisco, Napa Valley appeared on the wine charts in the mid-1800s: the first vines were planted in 1840, and the first commercial estate was established in 1858. Early on, climate and soil led Californian winemakers to favor grapes like Cabernet Sauvignon and to produce Bordeaux-like wine, with promising results: Napa's Inglenook estate won a gold medal at the world's fair in Paris as early as 1889.

Napa's true rise to fame occurred a century later, in 1976, during a blind test of French versus American wines, organized in Paris by the British wine merchant Steven Spurrier. This has gone down in history as the "Judgment of Paris." The combined votes of nine French judges crowned a Chardonnay from Château Montelena in Napa Valley as the best white wine and a Cabernet Sauvignon from Stag's Leap Wine Cellars, also in Napa Valley, as the best red.

The verdict shocked the wine community and boosted the reputation and economy of Napa Valley. Ten years later, in 1986, a rematch was organized between the same red vintages (1973) from the same estates — the French claiming that with aging, their wines would reclaim first place. Surprise hit again: first place went to yet another Napa Valley Cabernet Sauvignon, by Clos du Val, which had finished fifth in 1976 and evidently improved over time. The message became clear that France could no longer claim a monopoly on best wines and terroir.

Napa Valley's rise to prominence was the result of superior craftsmanship and superb terroir, in terms of both climate and soil. Interestingly, the prized estates of Château Montelena, Stag's Leap and Clos du Val lay astride both sedimentary and volcanic terrain. In fact, Napa Valley comprises an astounding diversity of soil types: lava flows and volcanic ash, marine limestone and marl, riverbed gravel, and clay, each contributing to the personality of the wines.

Napa Valley: A Tectonic Basin

A two-hour drive from San Francisco, Napa Valley runs parallel to the Pacific coastline, separated from the ocean by the Mayacamas Mountains and Sonoma Valley, a parallel array of hills and troughs that represents the tectonic border between the North American and Pacific plates.

California was built up over millions of years by the pasting together of chains of volcanic islands and stacks of sediments. Carried by the Pacific Farallon plate, these terranes were dumped against the growing continental margin like suitcases at the end of a conveyor belt. Around twenty million years ago, the head-on collision switched to a skirting, lateral motion, as the Pacific plate slid northwestward along California. With its margin now pulled apart rather than compressed, California developed a series of subsiding basins, parallel to the coast, separated by ridges. And so, Napa Valley came to be, flanked by mountain ranges.

Bending as it subsided, the floor of Napa Valley became a concave syncline: a folded-down, U-shaped sandwich of rock layers, with the youngest units lying at the center and the oldest ones — marine sediment and volcanic strata — sticking out at the periphery in the hillsides.

At its southern end, where it connects with San Pablo Bay, Napa Valley is 8 kilometers (5 mi.) wide at its mouth. Two main roads strike up the valley floor: Highway 29 on the western side and the more pastoral Silverado Trail to the east. The valley, 50 kilometers (30 mi.) long, becomes hotter and drier northward, away from the sea, and terminates in a dead end where its mountain flanks meet in an arcuate amphitheater dominated by volcanic Mount St. Helena — not to be mistaken with the active Mount St. Helens of Washington State.

Mount St. Helena, at 1,323 meters (4,342 ft.), is an uplifted pile of lava, like many other mountain crests lining Napa Valley and is over two million years old: its youthful, volcano-like shape is merely a fluke of erosion. Eruptions have long ceased in the area, but underlying pockets of magma are still hot enough to heat the water seeping into the ground. In Calistoga, half a kilometer down the road from Château Montelena, a small geyser, named Old Faithful, like its glorious analogue in Yellowstone Park, periodically expels hot water 20 meters (70 ft.) into the air. The Mayacamas range that separates Napa Valley from Sonoma Valley to the west harbors a much larger hydrothermal system.[2]

A Mosaic of Terroirs

Lava flows and ash layers in Napa Valley play a major role in the diversity and character of its terroir. The very notion of terroir — the expression of geography, geology and climate — was long downplayed in the United States. Wines were marketed more as a function of grape variety and brand name than emplacement. The irony is that Napa Valley is a remarkable showcase of terroir, with a multitude of soil types and microclimates.

Its winemakers now emphasize that Napa Valley comprises 33 main types of soil, further subdivided into a hundred or so subcategories. The valley is also home to three different climate zones, north to south, not to mention the many microclimates generated on its margins by the cold-air pockets rolling down the hill sides, and by local winds blowing through mountain clefts from adjacent valleys.

As a result, when the US Department of Agriculture created geographical appellations for American vineyards, named AVAs (American Viticultural Areas), Napa Valley was one of the first to be recognized, in 1981. It was further partitioned into sixteen subareas in 1989, starting with Stags Leap District:[3] a 400-hectare (1,000 acres) swath of vineyards on the eastern margin of the valley, below the basaltic ledge of the Vaca

Mountains. This was the first time in America that a viticultural area was established on the basis of terroir: a distinct soil and microclimate unit, highlighting in this particular case the basaltic rubble and the "solar oven" radiation of the rock face overlooking the vineyard.

Stag's Leap Wine Cellar, at the heart of this sub-AVA, is famous for claiming first place in the 1976 blind test of red wines in Paris. Besides the basaltic cliff radiating heat over the vines in the afternoon, the estate also benefits from a cool sea breeze in the evening, which funnels up the valley between the cliff and a row of parallel hills on the roadside. The temperature contrast between the day's heat and the night's coolness is an asset for the grapes, stretching out their ripening period and nicely balancing sugar and acidity. Moreover, there are two different soil types at Stag's Leap: an uphill basaltic rubble bringing "fire" to the Cabernet, according to estate founder Warren Winiarski, and a clay-rich sediment downslope bringing suppleness to the wine. Others attribute the tannic structure of the wine and its aromas of currant and blackberry to the volcanic component, and the red-berry, heady bouquet to the clay.

Diagonally across Napa Valley from Stag's Leap, in its northwestern corner, Diamond Mountain is another fine example of basaltic terroir. In this sub-AVA, lava flows are more recent (less than three million years old) and covered in places by a sprinkle of volcanic ash that blew in from the north. In 1968, Al Brounstein founded a vineyard in a small side valley encroaching the Mayacamas volcanic range that intersects three different soil types: a gravel terrace, a patch of soil cooked and oxidized by the lava (Red Rock Terrace), and a patch of gray ash erupted by the northern volcanoes (Volcanic Hill). Diamond Creek Vineyards covers 10 hectares (25 acres), crossed by a stream, with ponds and a beautiful landscaped garden.[4] Wine-tasting dinners are staged three days a week and include a tour of the grounds.

The winery overlooks Red Rock Terrace, sloping to the northwest. Underlain by red, lava-cooked soil, and rich in iron oxides, the lot is planted with fifty-year-old Cabernet Sauvignon. It yields the most classic, easiest-to-enjoy wine of the estate, with a bouquet of red currant, truffle, and tobacco, as well as flavors of cherry and blackberry, chewiness, and silky tannins.

Opposite the stream and facing south, Volcanic Hill is the warmest lot. Instead of baked clay, its soil is made up of gray volcanic ash, traced back to an eruption of Konocti volcano to the north, a few tens of thousands of years ago. The warm microclimate and powdery soil promote a more tannic version of Cabernet that ages particularly well, with a bouquet of red currant, a touch of violet and damp earth, truffle, leather, and often a whiff of graphite.

Finally, Gravelly Meadows occupies a dry streambed, where the stones efficiently drain away any excess moisture. Its wines have a bouquet that is both floral (violet and roses) and fruity (blueberry and blackberry), with jammy and earthy notes, ripe tannins, and a long, spicy finale.

Napa Valley has many other viticultural gems studded along its hill front, and many hold an important volcanic component. The most famous showcase of volcanic terroir in North America, however, lies outside California, 1,000 kilometers (620 mi.) to the north: Oregon's Willamette Valley.

Oregon's Great Lava Fields

Willamette is the last tributary of the Columbia River and joins it in Portland, 150 kilometers (100 mi.) short of the Pacific Ocean. Long before the dawn of civilization, the area was shaped by two monumental events, one volcanic, the other climatic.

The volcanic episode dates back to seventeen million years ago. At the time, California and Oregon rode the margin of a subduction zone, with an ocean plate diving under North America, and pods of magma rising through the strained crust. Something exceptional, however, happened much farther inland, on the border of today's Oregon and Idaho. Fissures many kilometers long and several meters wide poured out sheets of very hot, fluid lava that flowed toward the Pacific, across Oregon and Washington. Over two million years, flow after flow followed the same course to the sea, covering more than 150,000 square kilometers (58,000 sq. mi.), an area comparable to Washington State itself.

Large eruptions of the sort, known as flood basalts, are allegedly caused by the ascent of plumes of hot buoyant rock that rise from deep inside the Earth's mantle. Plumes are modeled as having an inverted teardrop shape, with the bulk of the magma erupting first, followed by a stretched-out tail.

No one was around seventeen million years ago to witness the fiery

Layers of lava from the Columbia River Basalts, cut into a cliff face by the Missoula catastrophic floods, in the Channeled Scablands of Washington State.

show as fountains of magma blew hundreds of meters into the air. At the base of the erupting fissures, fluid lava flowed downslope like motor oil, igniting and bringing down trees like matchsticks. Tens of meters thick, the lava front relentlessly proceeded westward, filling hollows and ramping over obstacles, covering 4 or 5 kilometers an hour (about 2.5 to 3 mi.), equivalent to a fast walking pace. At such a rate, the lava sheet probably took less than a week to travel from Idaho to the Pacific Ocean.

Overlapping each other, the lava flows that poured out in succession over two million years reached a cumulative thickness of 1 kilometer on average and 3 kilometers (nearly 2 mi.) in places — equivalent to the height of the Rocky Mountains — but creating no relief at the time, since the Earth's crust bent downward to accommodate the load. On the periphery of the main flow field, offshoots of lava spread into local depressions. One followed the axis of the future Willamette Valley southward, plastering a base layer of basalt in what would later become Oregon's finest terroir.

A Mighty Flood

The lava fields are still prominent in the landscape today, not only along the course of the Columbia River, which chiseled down through the pile and exposed them in great cliffs, but also farther north in Washington State. The landscape there is so bizarre that geologist J. Harlen Bretz conducted a survey in the 1920s and spotted a number of unusual features: wide valleys carved into the basaltic bedrock; vertical drops that looked like giant dry waterfalls; topographic swells up to 15 meters (45 ft.) high, as if some gigantic, undulating water flow had swept overland; teardrop-shaped mounds of gravel up to 500 meters (1,650 ft.) wide and several kilometers long; erratic boulders the size of houses standing in the middle of nowhere and a great distance downstream from their original location.

Pulling all the evidence together, Bretz suggested that a catastrophic flood once raced through the Pacific Northwest, carving the Columbia River Basalts into "channeled scablands" along the way. According to his theory, during the latest Ice Age, a glacial dam collapsed uphill in Idaho, releasing a volume of water close to that of Lake Michigan, which raced down through Washington and Oregon and emptied out into the Pacific Ocean.

When Bretz presented his audacious hypothesis in the 1920s and 1930s, he was met with sarcasm and disbelief: such a catastrophic concept was unacceptable in a science that embraced gradualism — the slow modification of landscapes over long periods — and considered catastrophism to be biblical folklore. Ironically, it took a trip to Mars to reopen the case: poring over satellite imagery of the red planet in the 1970s, geologists identified unmistakable signs of giant floods, characterized by the same carved channels, dry waterfalls, and streamlined islands that Bretz had signaled in the northwestern scablands. The latter were finally recognized as such: the result of catastrophic flooding on a gigantic scale.

There is a happy end to this story. In 1979, in the wake of the Martian study, Bretz was honored at age ninety-seven by the Geological Society of America for his pioneering work in the scablands: he received the prestigious Penrose Medal and a standing ovation.

Details of the Missoula Floods, as they came to be known, have now been filled in. Toward the end of the last Ice Age, fifteen thousand years ago, at a time when hunter-gatherers were entering North America via the

Bering Strait, offshoots of the ice cap that covered Canada stretched into Montana and Idaho, blocking rivers and damming meltwater to create a large glacial lake, measuring 7,700 square kilometers (3,300 sq. mi.) and 600 meters (1,900 ft.) average depth — half the volume of Lake Michigan.

The system reached a point at which the water pressure lifted the ice dam like a sluice gate, giving way to a surge of muddy water that raced downslope. The wave probably reached a height of 200 meters (660 ft.) and a speed of 100 kilometers per hour (60 mph), ripping the bedrock along its course, rolling chunks of basalt the size of houses, and reaching the Pacific Ocean in less than ten hours. The discharge rate was probably on the order of 20 million cubic meters per second, approximately one thousand times the flow rate of the Mississippi River.

The giant steps carved in the Columbia River Basalts are indeed what Bretz had claimed them to be: ancient waterfalls up to 5 kilometers (3 mi.) wide and 150 meters (500 ft.) in vertical drop: three times the width and height of Niagara Falls. Downstream, part of the flood swirled up into Oregon's Willamette Valley, dropping its mud load on the site of the future vineyards.

Cores drilled in the valley reveal up to forty layers of silt that reach an aggregate thickness of 10 meters (33 ft.). Hence, several dozen floods raced down the valley over time, the ice dam closing the sluice gate at the end of each event, and allowing the buildup of another lake full of glacial water, until it foundered again. On average, the great floods probably occurred twice a century, until the final retreat of glaciers out of Montana, in the waning stages of the Ice Age, 12,700 years ago.

First flooded with lava, then with mud, Willamette Valley thus experienced a tumultuous history, long before humans settled the land and planted vines in its multilayered soil.

Willamette Valley and Pinot Noir

Whereas Napa Valley is home to America's best Cabernet Sauvignon, challenging Bordeaux, Willamette Valley turns out some of the best Pinot Noir in the world, standing up to Burgundy.

This new challenge to French wines also began in Paris, three years after the competition that crowned Napa Valley. This time it was the Wine Olympics, organized in 1979 by the *Gault et Millau* food and wine guidebook. Burgundy came out the clear winner, but a Pinot Noir from Oregon's Eyrie Vineyards finished tenth, an encouraging result that did not go unnoticed. The following year, during a rematch in Burgundy's wine capital of Beaune, Eyrie's Pinot Noir clinched a surprising second place, behind a Burgundy Chambolle-Musigny. Intrigued by such a strong showing, competition organizer and wine mogul Joseph Drouhin and his daughter Véronique visited Willamette Valley in the 1980s and were quick to recognize the terroir's potential. By 1988 they had purchased land and planted their own vines in the valley, founding Drouhin Oregon.

Willamette Valley runs north-south between two mountain belts: the folded marine sediments of the Coastal Range to the west, separating the valley from the Pacific Ocean, and the volcanic Cascades to the east. Flowing northward, the Willamette River runs into the east-west Columbia River — the major axis that funneled lava flows and glacial floods down from Montana to build Pinot Noir's future terroir.

Oregon's prized vineyards were planted, starting in the 1960s, near the confluence of the two rivers, a rather surprising choice on account of the relatively cool climate and rainy fall season. The experience was undertaken on the grounds that the local climate was similar to Burgundy's, with the Cascades Range protecting the valley from the dry desert influence to the east, and the Coast Range downplaying the maritime dampness from the west. There is just enough of a breach — the Van Duzer Corridor — to let a welcome sea breeze blow in during the summer and cool the vineyards.

Contrary to Burgundy's straightforward limestone ridge, Willamette Valley's layout is rather complex, in terms of both geography and geology. Most estates line the many circumvolutions of the Coast Range that extends ridges and detached outliers into the valley. In most locations, one finds from the bottom up the Missoula Floods' loam-covered valley floor, layers of basalt, and piedmont marine sediments. The combination makes up a rich and complex terroir in which the influence of soil on wine can be directly appraised: one of the best areas in the world for conducting such research.

Willamette Valley contains twenty-three different soil types (versus thirty-three in Napa Valley), and in 2008 it was divided into six sub-AVAs: Chehalem Mountains, Dundee Hills, Eola–Amity Hills, Ribbon Ridge, Yamhill-Carlton District, and McMinnville.

Chehalem Mountains, the northernmost sub-AVA, is centered on a ridge of Columbia River Basalt, jutting out into the valley. The soil consists mainly of broken-down, altered lava rubble with a substantial amount of fine loam blown in from the nearby valley floor. Known as the Laurelwood series, this family of soils is the perfect illustration of Oregon's two catastrophic events: the Columbia River Basalts and the ice-age Missoula floods. Forming the ground of most of Chehalem Mountains sub-AVA and extending into adjacent areas, the Laurelwood series accounts for 330 hectares (85 acres) of vineyards: the third most planted soil type.

Driving past Chehalem Mountains, one reaches a second basalt ridge — Dundee Hills — framed by the Willamette River and its tributaries. This sub-AVA is underlain by the Jory soil series: a deep red soil that contains the largest fraction of basalt in the valley and accounts for 600 hectares (1,500 acres) of Pinot Noir in the Dundee Hills and elsewhere — the number-one soil type sought by winemakers. Dundee Hills' basaltic slope, facing southeast, is sliced into sections by several streams, much like Burgundy's Côte d'Or, which explains why it was one of the first areas settled by winemakers, starting with Eyrie Vineyard, Erath Winery, and later Drouhin Oregon.

Farther south, Eola–Amity Hills is the third volcanic sub-AVA in the valley (80 hectares, or 200 acres, of vines), circling around a basalt-capped outlier and blending the Jory soil series with a shallower component (the Nekia series) that contains coarse chunks of lava.

Between the basalt-capped ridges, as one proceeds upslope to the piedmont flanks of the Coast Range made up of sandstone and schist, two sub-AVAs promote the sedimentary, marine character of their terroir: Ribbon Ridge and Yamhill-Carlton District. Their soil is brown-colored, silica-rich and relatively deep at nearly 80 centimeters (3 ft.) down to bedrock: named the Willakenzie series, it is the second most planted soil type in the valley at 500 hectares (1,250 acres).

The sixth and last sub-AVA, McMinnville, downslope from the Van Duzer Corridor, owes its character as much to its marine soil as to the cool breeze that funnels down the cleft through the Coast Range.

There are also several hundred hectares of vines down in the flats, on the multilayered blanket of loam laid down by the Missoula Floods. While the loam's fertility is an asset for most crops, it is more of a drawback in viticulture: compounded with the flatness of the valley floor, the

Woodburn and Willamette soil series yield wines that are somewhat less complex than those from thinner soils and steeper slopes, and are often blended with them rather than raised and sold separately.

Ocean versus Lava: Pinot Noir Takes the Stand

In Burgundy, Pinot Noir is planted on marine limestone only, laced with variable amounts of clay, where it yields a remarkable suite of Premier Cru and Grand Cru wines, with subtle differences from one village to the next, and even within the same village and across adjacent lots.

Willamette Valley offers a broader comparison in the range of flavors and characters that Pinot Noir can take on, as it spans both marine sediment and lava. The comparison is all the more relevant because Pinot Noir is a "chameleon" grape, particularly well suited to express the subtleties of terroir, and both soil types—sedimentary and volcanic—lie side by side on many a slope, within the same climate, canceling out other factors that might influence the character of a wine, although the grape's clone variety, subtleties of microclimate, and differences in vinification still have a say in the end result.

In any case, major differences stand out between both types of terroir. On sedimentary, silica-rich Willakenzie soil, Pinot Noir yields a darker wine with a complex bouquet, both flowery and fruity, tones of cherry, blackberry, and sun-dried prune, as well as whiffs of lavender, cedar, licorice, tobacco, coffee, and leather. On basaltic, Jory-type soil, the wine ends up having a lighter shade of red, with aromas of red berries: cherry, raspberry, and red currant.

One explanation is that the sedimentary soil is better at evacuating excess water thanks to its lower clay content and the abundant quartz grains that loosen up its texture. The resultant hydric stress might concentrate tannins and promote the earthy, roasted, and gamy notes (coffee or leather) in the wine's bouquet. Volcanic soils, in contrast, retain some dampness because of their higher clay content (resulting from the breakdown of feldspar and glass in the basalt) and provide extra water to the grapes, encouraging fruitier aromas and more supple tannins.

These are of course first-order approximations. As is the case in Burgundy, slope, orientation, and altitude all play a role, as does the variety of Pinot Noir planted (Pommard or Dijon clone) and the style of vinification. Altitude also comes into play. At the lowest, hottest elevations, grapes ripen faster and accumulate more sugar, which later turns to alcohol. Higher up, the ripening phase is longer, giving grapes more time to develop complex aromatic molecules, which yield a greater range of scents and flavors and a finer, silkier texture. Finally, in the highest reaches of a vineyard, the cool climate raises the acidity of the grape juice, resulting in crisp wines with a fresh "bite."

Because of this complex interplay, Oregon winemakers often admit that they still cannot recognize in a blind test the soil series, or even the sub-AVA, of any given wine. They stress that vines are still too young in the Willamette Valley — half a century at most — to have extended their roots deep enough to reach bedrock and acquire their "adult" character, and that winemakers themselves have not yet had enough time to assimilate all the subtleties of their terroir.

Wine lovers are thus encouraged to travel to Oregon and carry out their own investigation. With over two hundred estates to visit in the Willamette Valley, and half a dozen wines to taste in each one, this is no small project.

Hawaii and Coffee

After all this wine, it is fitting to end this book with a cup of coffee. Besides their celebrated terroirs devoted to wine in California and Oregon, the United States also feature a fabulous volcanic terroir for coffee, in the Big Island of Hawaii, in the Pacific Ocean.

Hawaii's Big Island is a cluster of several volcanoes that grew in succession over a deep mantle hot spot, including Mauna Kea, Hualalai, Mauna Loa, and Kilauea — the latter two harboring eruptive activity today. The other major islands of the chain — Kauai, Oahu, Molokai, Lanai, and Maui — also bear prized agricultural belts on their now-extinct volcanoes, such as pineapple groves on Lanai, but coffee on the Big Island is the most celebrated, volcano-friendly crop.

The coffee plant originated in the Ethiopian Highlands of the volcanic East African Rift, before it was introduced to South and Central America, where it thrived in the warm tropical climate. Hawaii followed suit, when the Royal Hawaiian Court visited Europe in 1825 and fell in love with the beverage. Returning home by way of Rio de Janeiro, the Oahu chief Boki stocked up on Brazilian coffee saplings and planted them on his island but failed to obtain exploitable crops. When planted on the Kona coast of the Big Island a few years later, however, they did remarkably well, and production took off. The coffee plant had met its terroir, and despite several market crashes along the way, Kona coffee became one of Hawaii's flagship exports. It is ranked today as one of the best coffees in the world.

Finding the proper terroir for Hawaiian coffee, besides the positive influence of volcanic soil, has mostly to do with climate. Because of its varied topography, prevailing winds, and ensuing precipitation, Hawaii's Big Island possesses seven of the nine major climate zones defined on Earth, including hot desert, summer–dry cool temperate, and tropical monsoon. Within this wide range, it is the tropical winter–dry climate that best befits Kona coffee and, within it, a special microclimate that combines morning sunshine, shady afternoons, and cool evenings with moderate rainfall.

Consequently, the coffee belt stretches over 50 kilometers (30 mi.), from Kona airport to the southern tip of the island, along the western coast on the lower flanks of two overlapping volcanoes: Hualalai and Mauna Loa. Although the slopes face west, the shallow incline of the volcanic shields lets the morning sun glide over the plantations. By early afternoon, although the coast remains sunny, making Kona a popular tourist destination, a cloud cover forms upslope, cutting down the sun's rays and bringing the requisite coolness to the coffee plants. Precipitation remains low at 1,000 to 1,500 millimeters annually (40–60 in.) and can be supplemented with some irrigation.

Along the Kona coffee belt, most plantations are located at elevations between 230 and 750 meters (800–2,500 ft.), with the upper half of the range yielding the highest-quality coffee. The main reason for this is that afternoon cloud cover is more consistent at higher altitudes and nights are cooler: coffee beans grow slower as a result and become larger and denser. The cooler climate also enhances acidity and provides a "brighter cup," as well as rich sugars and a floral character, whereas lower elevations limit acidity and produce a softer, "mellow cup."

Besides climate, the basaltic "cinder" or scoria—hole-ridden chunks of lava—and its broken-down granular soil also contribute to the success of coffee in the Kona belt. Coffee plants need well-drained soil, which the porosity of scoria provides. On a Kona plantation, a volcanic substrate

WINES OF HAWAII

Hawaii is famous for coffee, tropical fruit, macadamia nuts, and honey. Growing grapes and making wine do not seem to fit into the cultural practices of the islands, but the experience was worth a try and first attempted on Maui: grapes were planted in 1974 on the southwestern slope of Haleakala volcano, at an elevation of 550 meters (1,800 ft.). As the pioneering winemakers waited for their grapes to grow, they experimented with pineapples and produced sparkling as well as still pineapple wine that met with great success. In 1984, they released their first grape sparkling wine from Carnelian—a cultivar created at the University of California in the 1970s by crossing Grenache, Carignan, and Cabernet Sauvignon. They have since branched out into a variety of cultivars, including Syrah, Grenache, and Malbec for the reds, and Viognier and Chenin Blanc for the whites, with the latter having replaced Carnelian in their sparking Blanc de Blanc. (Maui Wine, Ulupalakua Vineyards, 14815 Pilani Highway, Kula, HI 96790. Tel.: (808) 878 6058; www .mauiwine.com. Open 10 a.m.–5 p.m.; $14 wine-tasting fee).

The Big Island also has its own winery, started in 1986 on the slope of Mauna Loa, 1,200 meters (4,000 ft.) above sea level. Here too fruits other than grapes are fermented and used in original blends, including starfruit, *lilikoi* (passion fruit), papaya, guava, jaboticaba, and even macadamia-nut honey. At this Volcano Winery, located just north of the Kilauea caldera, the first grapes grown were Symphony—also created at the University of California, by crossing Muscat of Alexandria and Grenache Gris—until a fire in the year 2000 destroyed most of the vineyard. Since then, new French-American hybrids have been planted, such as Maréchal Foch, Chambourcin, and Cayuga white, as well as Pinot Noir and Pinot Grigio, with the belief that Pinot might adapt well to a terroir reminiscent of Oregon's Willamette Valley. It will be interesting to follow over the years the evolution of Volcano Winery's Pinot Noir, on a very alkaline soil that should amplify the grape's natural cherry and spicy bouquet. (Volcano Winery, 35 Pi'i Mauna Drive, Volcano, HI 96785. Tel.: (808) 967 7772; volcanowinery.com. Open 10 a.m.–5:30 p.m.; $5 wine-tasting fee.)

typically displays from top to bottom, in cross-section, less than 10 centimeters (2–3 in.) of organic soil, an equivalent thickness of volcanic cinder, and then hard lava. Brittle lava is generally cracked, which provides large-scale, extra permeability to absorb rainwater and often contains hollow lava tubes that serve as underground pipes to speed up the drainage job.

There is undoubtedly a chemical factor as well that comes into play with respect to Hawaii's volcanic terroir. When thinking about volcanoes, the first element that comes to mind is sulfur. It is present in high levels in basaltic soils, such as those of Hawaii, and is known indeed to play an important role in the flavor of coffee, promoting the formation of revered aromatic compounds during the roasting process, such as the "roasted aroma" itself, attributed to the sulfur-containing furfurylthiol molecule. In addition, although it does not enter their chemical formulas, sulfur plays a part in promoting the formation of citric and malic acid, responsible for aromas of orange blossom and green apple, respectively. But sulfur is only the tip of the iceberg: volcanic soil brings to coffee and other crops a cornucopia of essential chemical elements.

After wine and coffee, why not top off this review with a shot of rum? What a great excuse to visit the rich volcanic terroirs of Guadeloupe and Martinique, in the French Caribbean. But that's a whole other story . . .

Guide Section: Visiting California and Oregon

Napa Valley and Willamette Valley are conveniently located in adjacent states — California and Oregon — and are separated by a mere 1,000 kilometers (600 mi.), roughly a ten-hour drive. Spring, summer, and early fall are good times to visit; the late fall can sometimes bring heavy rainfalls to Oregon, although Willamette Valley is in the rain shadow of the Coast Range.

A Tour of Napa Valley (California)

Two hours by car from San Francisco, Napa Valley has become a popular tourist destination, with more than three million visitors a year. There is even a wine-tasting train that departs the city of Napa several times a day, with lunch or dinner on board, and a variety of tour options to visit vineyards along the way.[5]

There are two roads up the valley: Highway 29 to the west and the more pastoral Silverado Trail to the east, along the Vaca Mountains, described here.

Driving eastward out of Napa city, then north for about 8 kilometers (5 mi.), one first reaches a row of low hills jutting out of the valley floor. To the right is the entrance to Stag's Leap Wine Cellars:[6] the estate that placed first in the Judgment of Paris wine test in 1976. The estate's wines are on the expensive side ($110 to $225 a bottle), but one can taste them and visit the grounds for $50. The basalt cliffs overlooking the estate play an important role in the soil composition and microclimate (see page 161).

Back on the road (with a designated driver), one leaves the Stags Leap District sub-AVA for Oakville, then Rutherford—a terroir more sedimentary than volcanic, featuring gravel and loam. Volcanic bedrock is never far, however, and geology buffs should veer off Silverado Trail, a kilometer past the Mumm Napa Winery, onto Sage Canyon Road (Highway 128) toward Lake Hennessey. After a couple of kilometers, there is a sluice gate on the lakeside to the left; on the right side, the road bank is studded with blocks of green rock. These are remnants of the Pacific plate's volcanic seafloor, uplifted and pasted onto the Californian margin during a tectonic collision.

Back on Silverado Trail, 3 kilometers (2 mi.) to the north, one reaches the Saint Helena sub-AVA. Past Zinfandel Lane is a cream-colored road bank: a layer of pumice deposited by explosive eruptions, over three million years ago. The subsidence of Napa Valley, bent into a syncline, has tilted the initially horizontal ash layers into nearly vertical strata that break off into slabs on the roadside.

A couple of kilometers farther, at the Dear Park Road intersection, Silverado Trail crosses an alluvial fan: a sheet of sand and gravel dumped at the bottom of the slope by an ancient stream (most of it is covered by vegetation). Winemakers are particularly fond of alluvial fans (known as *combes* in Burgundy) for their coarse, well-drained soil. There are a dozen such fans along Napa Valley's mountain fronts.

Opposite Duckhorn Vineyards, the road embankment cuts through cream-colored mixed ash and lava beds. Belonging to a volcanic hill known as Glass Mountain, these silica-rich strata contain large chunks of black and shiny obsidian (a glassy form of rhyolite).

Another 2.5 kilometers (1.5 mi.) down the road, in a left-hand curve, 150 meters (500 ft.) past the entrance of Rombauer Vineyards, notice the brick-red outcrop in the embankment: the contact between a hot ash flow (cream-colored pumice) and the sediments that it baked and oxidized into red rock. In the pumice layer, one can also spot shiny chunks of obsidian.

Heated up by all these volcanic discoveries, one might want to back-track to Rombauer Vineyards (visit by appointment only; make sure to call ahead of time — wine tasting is $25 per person, or $100 with a guided tour).[7] The estate owns vines in different areas of Napa Valley and show-cases Cabernet Sauvignon from Saint Helena and Atlas Peak; it also mar-kets Zinfandel, Merlot, and Chardonnay — some from Napa, others from elsewhere in California.

Resuming the drive, up Silverado Trail, one reaches the northern end of Napa Valley and its Calistoga sub-AVA, framed by the high-standing Howell Mountain sub-AVA to the east and Diamond Mountain sub-AVA to the west. Straight ahead, Saint Helena's volcanic hill acts as the valley's head wall. Off Tubbs Lane, nested in an oasis of trees, stands Château Montelena, the estate that upset Burgundy whites in the 1976 Judgment of Paris.[8] Here one can opt for a winery tour (Monday through Friday, 10 a.m., $50 per person), and a small ($30) or expanded ($60) wine-tasting session of Cabernet Sauvignon, Chardonnay, and Zinfandel.

This is a great way to end a tour of Napa Valley — or at least the first day of it, since one might also want to visit the western side of the valley upon returning to San Francisco by way of Saint Helena Highway (Route 29). And for those geology buffs who want an extra thrill, upon exiting Château Montelena and continuing 800 meters (about half a mile) down Tubbs Lane, a small geyser — pompously named Old Faithful in reference to its Yellowstone big brother — performs every fifteen minutes, with a 20-meter (66 ft.) spray of hot water ($15 entrance fee).

A Tour of Willamette Valley (Oregon)

The city of Portland is the entrance gateway to Oregon's Willamette Val-ley. The region is clearly volcanic, as one can tell from the looming volca-noes of the Cascade Range to the east. Only 80 kilometers (50 mi.) distant,

Mount Hood last erupted in 1866. Towering 3,400 meters (10,000 ft.) over Oregon's forests, the volcano is well worth a hike, and its many trails are open from May to October. Several ski runs are open on the volcano during winter.

The Cascade Range affected the terroir much less than the Columbia River Basalts that erupted 1,000 kilometers (620 mi.) to the east. To have a look at their spectacular makeup, one should drive up the Columbia River Highway (Highway 84), east out of Portland, for an hour or so. About 30 kilometers (19 mi.) up the valley, good views of the basaltic cliffs open up. Take the Corbett exit and follow the crest road — East Historic Columbia River Highway — to Crown Point Vista House for a panoramic view of the valley.

Past Crown Point, the historic highway proceeds downslope, crossing a gorge that displays a waterfall cascading down a wall of columnar-jointed basalt (Latourell Falls). Before reaching the valley, one passes another cascade (Multnomah Falls) with a stone pedestrian bridge spanning the gorge.

To visit Willamette Valley and its celebrated vineyards of Pinot Noir, head south out of Portland. Branching off Interstate 5, take the Pacific Highway West (Route 99W) that crosses the hills toward Sherwood and Newberg. These are the Chehalem Mountains, one of Willamette's main sub-AVAs that features volcanic soil. Upon reaching Newberg, one then penetrates the sub-AVA of Dundee Hills, also basaltic in nature.

After crossing the city of Dundee, take SW Ninth Street to the right, which changes names to NE Worden Hill Road after a kilometer or so. After about 2 kilometers (1 mi.) through vineyards and a few curves up a hill, there is a turnoff to the left, NE Knudsen Lane, that leads to Erath Winery.[9]

One of the first estates established in the valley, in 1969, Erath offers a wine-tasting session for $30, highlighting Pinot Noir grown in the volcanic Jory soil series, as well as wines from several other types of soil.

Back on route 99W, drive 20 kilometers (12 mi.) farther south to Lafayette, then to McMinnville. There is an interesting geological side trip here, as the road intersects Route 18 (Salmon River Highway). Take a right on this route and drive 10 kilometers (6 mi.) toward the southwest: a road sign signals a hiking trail, and slightly past it, a drivable road (SW Oldsville Road) that leads to Erratic Rock State National Site.

The site symbolizes what the Missoula Floods were all about: perched on a hilltop lies an "erratic rock," weighing thirty-six tons, that originated in Canada. It was first transported south to Montana on a glacier that ended up damming Lake Missoula. The rock was then swept downstream when the dam broke loose, encased in a frame of ice that acted as a floater. About 1,000 kilometers (600 mi.) down the Columbia River Valley, as the current backed up into Willamette Valley and the flood subsided, the rock was left stranded on the hill.

Back on Salmon River Highway, head for Amity and the Eola–Amity Hills sub-AVA, another vine-covered mountain block. After driving through Amity, continue southward on 99W for about 7 kilometers (4 mi.), take a left on Lincoln Zena Road that runs across fields and then ramps up into woodland. After 5 kilometers (3 mi.), take a left on Bethel Heights Road and drive through the vines to the entrance of Bethel Heights Vineyard.[10]

From the wine-tasting terrace, enjoy a magnificent view over the estate and sample a selection of Pinot Noir wines from volcanic lots (Jory soil series), which you can compare with their marine sediment counterpart, from much farther downslope. This contrast of wines and soil types is what Willamette Valley is all about. Reflect on the amazing story of fire and ice that made this terroir what it is as you sip your Pinot Noir and watch the sun set over the hills.

Notes

Chapter 1

1   Riggs Fulmer, "Then and Now," *Oregon Wine Press*, 2009, https://www.oregonwinepress .com/article?articleTitle=then-now--1262800152--189--news.

2   Andisol (spelled *Andosol* outside the United States) is named not after the Andes volcanoes but after the Japanese words *an do*, meaning "dark soil."

Chapter 2

1   The name Vinsanto leads to some confusion because there is another *vin santo* ("holy wine"), produced in northern Italy and similar in style that owes its name to the fact that it is vinified during Holy Week. Spelled as one word, Vinsanto refers to "wine from Santorini": the abbreviation "Santo" was written on the barrels shipped out of the island.

2   It so happens that *caldera* is the word in Spanish and Portuguese for "cauldron," on account of its shape.

3   Check opening days and hours, often at odds with published information.

4   Open every day from 8 a.m. to 8 p.m. during the high season (April–October), with an early closing time (3 p.m.) on Thursdays. In the off-season from 8 a.m. to 3 p.m. (closed on Monday). Tel.: (30) 22860 81939.

5   Gaia Wines is open 10 a.m.–6 p.m., late April through late October. It is better to contact the estate to check opening hours and inform them of your visit. Tel.: (30) 22860 34186; www.gaiawines.gr/santorini-winery-eng/. Wine tasting ranges from approximately €10 for four wines to €20 for eight wines.

6   In the quieter backstreets of the village, Glaros Hotel is a good address. Tel.: (30) 22860 32601; www.glaros-santorini.com. Co-owner Stavros Pelekis began his career as a winemaker at the Boutari estate. His restaurant proposes a good, well-priced wine list and menu.

7   Hatzidakis Winery is visited by appointment only. Tel.: (30) 69700 13556; www.hatzi dakiswines.gr/en.

8   Domaine Sigalas. Tel.: (30) 22860 71644; www.sigalas-wine.com/english/. Wine by the glass, or full two-hour tour, with ten wines and food pairings (around €90).

9   Boutari. Tel.: (30) 22860 81011; www.boutariwinerysantorini.gr. Daily forty-five-minute wine tour at 11 a.m., including the tasting of five wines (around €20); and more intricate, ninety-minute tours in the afternoon at 1, 2, and 3 p.m. (around €40).

10 Gavalas Winery. Tel.: (30) 22860 82552; www.gavalaswines.gr/en. Open 10 a.m.–8 p.m. in the summer (check hours during other seasons).

11 Santo Wines. Tel.: 30 22860 28058; www.santowines.gr/en. Cellar tour €11; wine by the glass, and wine tastings from €20 to €45 (sixteen-wine flight).

12 Volcan Wines (Wine Museum). Tel.: 30 22860 31322; www.santoriniwinemuseum.com. Opening hours: 10 a.m.–7 p.m. (April–October); closes at 5 p.m. in the off-season. Wine tour (museum and wine tasting): €10.

13 Artemis Karamolegos Winery. Tel.: (30) 22860 33395; www.artemiskaramolegos-winery .com/en/. Open April through October (11 a.m.–8 p.m.). Winery tour and wine tasting (around €10).

14 Estate Argyros. Tel.: (30) 22860 31489; www.estate-argyros.com. Wine tasting: €15; wine tasting and tour of vineyards: €40.

15 Canava Roussos. Tel.: (30) 210 9400017; www.canavaroussos.gr/uk/company.html. Wine by the glass.

Chapter 3

1   DOC stands for *Denominazione di Origine Controllata* (Controlled Appellation of Origin).

2   Pliny the Elder, *Naturalis Historia* 3.9, translation by the author. For clarity, I added "wine and wheat," Pliny substituting for each produce the name of its god.

3   By means of comparison, a Pompeiian workman earned four sesterces (sixteen asses) a day and probably kept to house wine. A senator, however, earned over a million sesterces a year and could certainly afford the Salerno.

4   Strabo, *Geography* 5.4, translation from Loeb Classical Library edition, 1923, retrieved from http://penelope.uchicago.edu/Thayer/E/Roman/Texts/Strabo/5D*.html.

5   Pliny the Younger, *Letters*, LXV, to Tacitus, translation by Harvard Classics, 1909–1914, simplified by the author.

6   Pliny the Younger, *Letters*, LXVI, to Tacitus, translation by Harvard Classics, 1909–1914.

7   In the 1980s, the skeletons of three hundred villagers were discovered in boathouses on the outskirts of town, where they were waiting out the eruption and were caught by the deadly surge.

8   In a corner of Foro Boario, a small shed was unearthed that stored eight large clay jars (*dolia*) buried flush with the ground. Roman winemakers were well aware that burying the vats in cool soil kept the temperature down and slowed down fermentation, thus improving the wine.

9   IGT stands for *Indicazione Geografica Tipica*, or Protected Geographical Indication.

10 Lucius Columella, *Res Rustica* 3.2, translation by Loeb Classical Library, 1941, retrieved from http://penelope.uchicago.edu/Thayer/e/roman/texts/columella/home.html.

11 Pliny the Elder, *Naturalis Historia* 14.8 and 14.5, translation by the author.

12   Lucius Columella, *Res Rustica* 3.21, Loeb Classical Library edition, 1941, retrieved from
      http://penelope.uchicago.edu/Thayer/e/roman/texts/columella/home.html.
13   Most notably at Lake Nyos, Cameroon, in 1986.
14   Besides its use as a natural sweetener, sorbitol is used in cardiovascular medication, as a
      coronary vasodilator.
15   Cantina del Vesuvio, via Tirone della Guardia 18, Trecase. Tel.: (39) 081 536 90 41; www
      .cantinadelvesuvio.it. Wine tasting every day 11 a.m. to 3 p.m. (until 4 p.m. on week-
      ends): includes lunch for approximately €30.
16   Azienda Sorrentino, via Frusco 2, Boscotrecase. Tel.: (39) 338 135 44 15; www.sorrentino
      vini.com/en/. Wine-tasting tour every day 10 a.m.–3 p.m.: approximately €10–€20 per
      person, €30–€40 with lunch.
17   Cantine Olivella, via Zazzera 14, Sant'Anastasia. Tel.: 39 081 531 13 88; www.cantine
      olivella.com.
18   Azienda Fuocomuorto, contrada Croce dei Monti, 22, Ercolano; Tel.: (39) 081 739 46 55
      or (39) 331 522 69 37; www.fuocomuorto.it. Also the B&B page: www.bedandbreakfast
      vesuvio.com. For a double room with breakfast, plan on approximately €60 per night.

Chapter 4

1   Lava fountains at Mount Etna usually last a couple of hours. There were sixty-six such
     events in 2000 (more than one a week) and twenty-five in 2011–2012 (two a month).
     Taking into account lulls between such series, the probability of witnessing a lava foun-
     tain during a three-hour trek to the summit plateau is on the order of 1 in 500.
2   Saint Agatha's feast day is celebrated on February 5 — the day of her death in 251 — when
     her relics are carried outside Catania's cathedral. There is also a celebration on August 17,
     the day her relics were returned from Constantinople to Catania in 1126.
3   DOP stands for *Denominazione di Origine Protetta*, Protected Designation of Origin,
     a label certifying that the produce is locally grown and packaged, using traditional
     methods.
4   Etna vineyards share this record of 1,200 meters (3,937 ft.) with the alpine Swiss and
     Italian vineyards of Visperterminen and Morgex. But it is the Canary Islands, in the
     Atlantic Ocean, that claim the overall European record at 1,650 meters (5,400 ft.). The
     world record belongs to Argentina at 3,100 meters (10,400 ft.).
5   Strabo, *Geography* 6.2, retrieved from http://penelope.uchicago.edu/Thayer/e/roman
     /texts/strabo/home.html.
6   IGT stands for *Indicazione Geografica Tipica*, Protected Geographical Indication; DOC
     stands for *Denominazione di Origine Controllata*, Registered Designation of Origin.
7   Although the Rhône Valley's Hermitage is made up of granite and gravel, and Mount
     Etna's Monte Colla features limestone, the idea is that both hills are made up of differ-
     ent rocks than their surroundings.
8   Trachybasalt is a form of basalt slightly enriched in silica and alkali metals (sodium and
     potassium).
9   *Guida alle Cantine dell'Etna* (2008); www.ct.camcom.gov.it/documenti/vini/guida
     _cantine_etna.pdf. Also, the Etna Wine Guide (2017); €10, plus €10 shipping.
10   Nicosia, via Luigi Capuana 65, Trecastagni. Tel.: (39) 095 780 67 67; www.cantine

nicosia.it/en/. Open for wine tasting Monday through Friday, 9 a.m.–noon and 3 p.m.–6 p.m.

11 Benanti, via Giuseppe Garibaldi 361, Viagrande. Tel.: (39) 095 789 09 28; www.vinicola benanti.it.

12 Scamacca del Murgo, via Zafferana 13, Santa Venerina. Tel.: (39) 095 950 520; www .murgo.it.

13 Tenuta San Michele, same address and phone as Scamacca del Murgo; www.tenuta sanmichele.it.

14 Scilio, Contrada Valle Galfina, Linguaglossa. Tel.: (39) 095 932 822; https://scilio.it/en/.

15 Gambino Vini, contrada Petto Dragone. Tel.: (39) 348 822 01 30; www.vinigambino.it /?lang=en.

16 Azienda Frank Cornelissen, via Canonico Zumbo 1, Passopisciaro; www.frankcornelis sen.it. For wine tours: pr@frankcornelissen.it.

17 I Vigneri, Palmento Caselle, via Abate 3, Milo. Tel.: (39) 366 662 25 91; www.ivigneri.it. For information regarding wine-tasting activities: info@ivigneri.it.

18 Vini Franchetti, contrada Guardiola, Passopisciaro. Tel.: (39) 094 239 54 49; www.vini franchetti.com/passopisciaro/. For a wine-tasting appointment: passopisciaro@vini franchetti.com.

Chapter 5

1 Mount Etna also claims Hephaestus and his Cyclops as underground residents.

2 IGT stands for *Indicatione Geographica Tipica*, a broad geographical denomination, whereas DOC, *Denominazione di Origine Controllata*, implies a more precise location and stricter quality requirements.

3 Tenuta di Castellaro, via Caolino, Lipari. Tel.: (39) 345 434 2755; www.tenutadi castellaro.it/en/.

4 Contrary to most other volcanoes of the Aeolian Islands, Stromboli erupts a basaltic to trachybasaltic magma, low in silica and relatively fluid.

5 The more recent name of Salina stems from the salt marshes developed on the southern tip of the island, facing Vulcano.

6 Paola Lantieri, Punta dell'Ufala winery. Tel.: (39) 336 712 905; www.facebook.com /MalvasiaLantieri/.

7 Tasca d'Almerita, Tenuta Capofaro, via Faro 3, Salina. Tel.: (39) 090 98 44 330/1; www .tascadalmerita.it/tenuta/capofaro/ and www.capofaro.it.

8 Carlo Hauner Azienda Agricola, via Umberto I, Lingua, Marina Salina. Tel.: (39) 090 984 31 41; www.hauner.it.

Chapter 6

1 The German word *Graben* is the scientific term for the dropped central axis of a rift zone. It is often framed on either side by an uplifted horst (also German), or "rift shoulder."

2 AOC stands for *Appellation d'Origine Contrôlée*, the protected designation of origin, bestowed to quality wines in France.

3 "Village" status is a special tier of the French classification scheme for wines used when,

within a region, the wines of a specific village stand out. The name of the village is then added to the title of the appellation or even supersedes it. For instance, within the region of Burgundy (standard appellation Burgundy), the wines of the village of Pommard are so distinguished that the appellation becomes Pommard — that of the village. In the region of Bordeaux, the village of Margaux is similarly honored for its wines, and so Margaux supersedes Bordeaux on the label, and so on. In Auvergne, the name of the distinguished village is (more modestly) added to the Côtes-d'Auvergne title rather than superseding it, such as Côtes-d'Auvergne-Châteaugay.

4 Cited in *Le Vignoble des Côtes d'Auvergne* (2013), translated by the author.

5 Another valley slightly to the south, Couze Pavin, offers a great view of the pyroclastic flows, eroded back into spectacular cliffs near Perrier village. Troglodyte dwellings were built into the cliff face and occupied until the mid-twentieth century. Near Saint-Diéry the caves form an intricate underground village that dates back to the Middle Ages.

6 Vulcania, route de Mazayes, 63230 Saint Ours les Roches. Tel.: (33) 4 73 19 70 00; www .vulcania.com/en/. Approximately €30 per adult, €20 per child. Open from late March to early November.

7 Volcan de Lemptégy, 31 route des Puys, 63230 Saint Ours les Roches. Tel.: 04 73 62 23 25; www.auvergne-volcan.com. Entrance fee: €15 per adult, €12 per child. Open from mid-February through late October (before April, open only Tuesday through Thursday).

8 Panoramique des Dômes (Puy de Dôme rack railway), La Font de l'Arbre, 63870 Orcines. Tel.: (33) 4 73 28 70 00; www.panoramiquedesdomes.fr. Open all year, hourly departures (every twenty minutes during summer). Approximately €15 per adult, €8 per child (round trip).

9 Grotte de la pierre de Volvic, 2 route du pont Jany, 63530 Volvic. Tel.: (33) 4 73 33 56 92; www.maisondelapierre-volvic.com/en/. Open from mid-February through late October (before April, open only Tuesday through Thursday; in September and October, closed on Saturday). Approximately €8 per adult, €6 per child.

10 No Control estate, Vincent Marie winemaker; wine shop: 3 rue de la Croix Badière, 63530 Volvic. Tel.: 33 6 64 23 50 22; www.vin-nocontrol.fr/en/. Visits and wine tasting by appointment.

11 Domaine Stéphane Bonjean, 88 rue du Clos, 63112 Blanzat. Tel.: (33) 6 83 12 88 90. Wine tasting by appointment.

12 Gilles Persilier, 3 bis rue du Centurion, 63670 La Roche Blanche. Tel.: (33) 4 73 79 44 42 (land), and (33) 6 77 74 43 53 (cell); www.persilier-vins.com.

13 Vieilles Vignes: made with old vines (typically over fifty years old).

14 Christian Liévain, 3 rue Haute, 63730 Corent. Tel.: (33) 6 85 69 28 04; email: christian .lievain@orange.fr

15 Cave Saint-Verny, 2 route d'Issoire, 63960 Veyre-Monton; Tel.: (33) 4 73 69 92 87; www .saint-verny.com. Open Tuesday through Saturday 9 a.m.–12 :30 p.m. and 2 p.m.–7 p.m.

16 Domaine Miolanne, D978 road between Plauzat and Champeix. Tel.: (33) 6 72 41 22 56; www.domainemiolanne.com. Open Monday–Saturday (10 a.m.–12:30 p.m. and 3 p.m.–6:30 p.m.).

Chapter 7

1   An arrangement was reached by the feuding colonies in 1479, granting the Canary Islands to Spain and Madeira to Portugal. The Canary Islands also had a native population, the Guanches, of North African Berber origin: they fought the European settlers and were decimated between 1478 and 1496.

2   Listán Negro was also planted in California by Spanish missionaries, hence its American name Mission Grape.

3   Casa del Vino La Baranda, autopista general del Norte, km 21. Tel.: (34) 922 57 25 35; www.casadelvinotenerife.com. A thirty-minute drive from Santa Cruz, via the TF5 highway: El Sauzal exit; at the roundabout, Calle Urbanizacíon; then after 200 meters (650 ft.), right turn into Calle San Simon.

4   El Grifo winery, LZ 30, km 11, San Bartolomé. Tel.: (34) 928 524 03; www.elgrifo.com /en/.

5   El Chupadero restaurant, La Geria, 3, Yaiza. Tel.: (34) 928 177 365.

6   La Geria bodega, carretera La Geria, km 19, Yaiza. Tel.: (34) 928 173 178; www.lageria .com/en/.

7   Bodegas Rubicon, carretera Teguise-Yaiza, 2, La Geria. Tel.: (34) 928 173 708; www .bodegasrubicon.com.

Chapter 8

1   The word *trap* means "staircase" in Norwegian and Swedish, illustrating the steplike pattern in which the basaltic layers erode along their margins.

2   A magma lens some 7 kilometers (4 mi.) below the surface heats an aquifer tapped by 350 wells providing steam to power plants that produce two thousand megawatts of electricity: enough to satisfy half the needs of California's northern counties, from San Francisco to the Oregon border.

3   Note the lack of apostrophe in the word *Stags* for the district. A Native American legend tells the story of a deer that escaped a group of hunters by jumping off a cliff. Stag's Leap Cellars, below the cliff front, was named after the legend. A second estate, founded nearby, also wished to use the name and did so by placing the apostrophe after the *s*, calling itself Stags' Leap. When it came to naming the sub-AVA, so as not to favor one estate over the other, the committee settled on Stags Leap with no apostrophe.

4   Diamond Creek Vineyards, 1500 Diamond Mountain Road, Calistoga, CA 94515. Tel.: (707) 942 6926; www.diamondcreekvineyards.com.

5   Napa Valley Wine Train. Tel.: (800) 427 4124; www.winetrain.com. Prices range from $150 to $350 per person.

6   Stag's Leap Wine Cellar, 5766 Silverado Trail, Napa, CA 94558. Tel.: (707) 261 6410; www.cask23.com. Open daily 10 a.m.–4:30 p.m. Tastings and tours range from $50 to $175 per person.

7   Rombauer Vineyards, 3522 Silverado Trail, Saint Helena, CA 94574. Tel.: (707) 963 6606; www.rombauer.com. Open 10 a.m.–6 p.m. by appointment only. Wine tasting: $25 per person, $100 with a guided tour.

8   Château Montelena Winery, 1429 Tubbs Lane, Calistoga, CA 94515. Tel.: (707) 942

5105; www.montelena.com. Walk-in wine tasting 9:30 a.m.–4 p.m. daily, $30 to $60, and 10 a.m. winery tour ($50).

9   Erath Winery, 9409 NE Worden Hill Road, Dundee, OR 94515. Tel.: (503) 538 3318; www.erath.com. Open 11 a.m.–5 p.m.

10   Bethel Heights Vineyard, 6060, Bethel Heights Road NW, Salem, OR 97304. Tel.: (503) 581 0943; www.bethelheights.com. Open 11 a.m.–5 p.m.; $15 tasting fee.

Bibliography

Carrera, F., and F. Latteri, eds. 2016. *Guida ai vini dell'Etna* [Etna wine guide]. Palermo: Cronache di Gusto. Published in 2018 in a bilingual Italian-English edition.

Couderc, D., ed. 2013. *Le Vignoble des Côtes d'Auvergne*. Aurillac: Éditions Quelque part sur terre.

Foti, S. 2012. *Etna: I vini del vulcano*. 2nd ed. Catania: G. Maimone Editore.

Krafft, M., and F. D. De Larouzière. 1999. *Guide des volcans d'Europe et des Canaries*. Paris: Delachaux et Niestlé.

Lopes, R. 2005. *The Volcano Adventure Guide*. Cambridge: Cambridge University Press.

Poux, M., ed. 2011. *Corent: Voyage au cœur d'une ville gauloise*. Paris: Errance.

Richet, P. 2003. *Guide des volcans de France*. Paris: BRGM éditions and Belin.

Tanneau, J., ed. 2004. *La Vigne et le vin en Auvergne*. Clermont-Ferrand: Chamina.

Vougioukalakis, G. 2006. *The Minoan Eruption of the Thera Volcano and the Aegean World*. Thira: Society for the Promotion of Studies on Prehistoric Thera.

Websites

Santorini

Artemis Karamolegos winery, Exo Gonia (English): www.artemiskaramolegos-winery
.com/en/
Boutari Winery, Megalochori (English): www.boutariwinerysantorini.gr
Canava Roussos estate, Episkopi Gonia (English): www.canavaroussos.gr/uk/motions
.html
Domaine Sigalas estate, Oia (English): www.sigalas-wine.com/english/
Estate Argyros, Episkopi Gonia (English): www.estate-argyros.com
Gaia Wines estate, Kamari (English): www.gaiawines.gr/santorini-winery-eng/
Gavalas Winery, Megalochori (English): www.gavalaswines.gr/en
Glaros Hotel, Kamari (English), www.glaros-santorini.com
Hatzidakis Winery, Pyrgos (English): www.hatzidakiswines.gr/en
Santo Wines, Pyrgos (English): www.santowines.gr/en
Volcan Wines and Wine Museum, Vothonas (English): www.santoriniwinemuseum
.com

Mount Vesuvius

Azienda Fuocomuorto estate and B&B, Ercolano (English): www.fuocomuorto.it and
www.bedandbreakfastvesuvio.com
Azienda Sorrentino estate, Boscotrecase (English): www.sorrentinovini.com/en/
Cantina del Vesuvio estate, Trecase (English): www.cantinadelvesuvio.it
Cantina Olivella estate, Sant'Anastasia (Italian): www.cantineolivella.com

Mount Etna

Azienda Frank Cornelissen estate (English): www.frankcornelissen.it
Benanti estate, Viagrande (English): http://vinicolabenanti.it/en/home/

Gambino Vini estate, (English): www.vinigambino.it/?lang=en
Guidebook of Etna estates (Italian): www.ct.camcom.gov.it/documenti/vini/guida
_cantine_etna.pdf
I Vigneri estates, Milo (English): www.ivigneri.it
Nicosia estate, Trecastagni (English): www.cantinenicosia.it/en/
Scamacca del Murgo estate and B&B, Santa Venerina (English): www.murgo.it and
www.tenutasanmichele.it
Scilio estate, Linguaglossa (English): https://scilio.it/en/
Vini Franchetti, Passopisciaro estate, Passopisciaro (English): www.vinifranchetti
.com/passopisciaro/

The Aeolian Islands

Carlo Hauner Azienda Agricola estate, Lingua, Salina (English): www.hauner.it
Paola Lantieri estate, Punta dell'Ufala, Vulcano (English): www.facebook.com
/MalvasiaLantieri/
Tasca d'Almerita and Tenuta Capofaro estates and hotel, Salina (English): www
.tascadalmerita.it/tenuta/capofaro/ and www.capofaro.it
Tenuta di Castellaro estate, Quattropani, Lipari (English): www.tenutadicastellaro
.it/en/

France's Hidden Volcanoes

Château Beaulieu and Villa Baulieu estates (English): www.chateaubeaulieu.fr/en and
www.villabaulieu.com
Domaine Miolanne estate, Neschers (French): www.domainemiolanne.com
Gilles Persilier estate, Gergovie (French): www.persilier-vins.com
No Control estate, Volvic (English): www.vin-nocontrol.fr/en/
Puy de Dôme rack railway (French): www.panoramiquedesdomes.fr
Saint-Verny wine cooperative, Veyre-Monton (French): www.saint-verny.com
Volcan de Lemptégy theme park (English): www.auvergne-volcan.com/visiting
-volcan-de-lemptegy
Volvic stone quarry museum (English): http://www.maisondelapierre-volvic.com
Vulcania theme park (English): www.vulcania.com/en/

The Canary Islands

Azores wines (English): www.picowines.net
Bodega El Grifo estate, San Bartolomé, Lanzarote (English): www.elgrifo.com/en/
Bodega La Geria estate, Lanzarote (English): www.lageria.com/en/
Bodega Los Bermejos estate, Lanzarote (English): www.losbermejos.com
Bodega Stratus estate, Lanzarote (English): www.stratvs.com
Bodegas Rubicón estate, Teguise-Yaiza, Lanzarote (English): www.bodegasrubicon
.com
Casa del Vino, La Barranda estate, Tenerife (English): www.casadelvinotenerife.com

California, Oregon, and Hawaii

Bethel Heights Winery, Salem, Willamette Valley (English): www.bethelheights.com
Château Montelena estate, Calistoga, Napa Valley (English): www.montelena.com
Diamond Creek Vineyards, Calistoga, Napa Valley (English): www.diamondcreek
vineyards.com
Erath Winery estate, Dundee, Willamette Valley (English): www.erath.com
Maui Wine estate, Maui, Hawaii (English): www.mauiwine.com
Napa Valley Wine Train (English): www.winetrain.com
Rombauer Vineyards estate, Saint Helena, Napa Valley (English): www.rombauer.com
Stags' Leap Cellar estate, Napa, Napa Valley (English): www.cask23.com
Volcano Winery estate, Big Island, Hawaii (English): www.volcanowinery.com

Credits

Figures

Plates

General Index

References to figures are denoted by an italic "f" following the page number; "pl" denotes numbered color plates.

acidity: of coffee, 170; of grape juice and wine, 6, 9, 20–21, 24–26, 81–82, 94, 117, 126, 134–35, 150, 152, 161, 169
Aeolus, 108
Agatha, Saint, 70, 179n2
alborello pruning, 90
alluvial fan, 173
almond, 67, 88, 126
altitude, and climate, 2, 8, 37, 77, 87, 92–94, 126, 168. *See also* elevation
aluminum, 2, 4, 24, 50
American Viticultural Area (AVA), 160
amphora, 14, 45–47, 61, 127–28, 133, 141
Andisol, 7, 177n1
anthocyanin, 9, 75, 81
AOC appellation, 131, 180n2
AOVDQS appellation, 131
Apollinaire, Sidoine, 128
apricot, 2, 37, 50, 57–58, 63, 116; liquor, 63
Arancello, 88
arenado, 147, 149
Arverni, 127
ash, volcanic, 2, 6–7, 15–16, 18, 23–24, 41–44, 49, 54, 59, 69, 76–77, 79, 121, 123, 125, 135–36, 139, 132, 144, 146, 147f, 159–61, 171, 173–74, pl16; ash cloud, column, fall, plume, 5, 15–16, 30, 42–44, 54, 114–15, pl31, pl41, pl52–53. *See also* cinder cone; pumice; pyroclastic flow

AVA (American Viticultural Area), 160
Avernus, Lake, 51–52

Bacchus, 40, pl12
banana, 146, 150
basalt, 9, 68, 76–77, 93–94, 97–98, 125–26, 128, 132–34, 132f, 139–41, 144–47, 150, 157–58, 160–61, 165–68, 171, 173, 175
base surge, 16, 30, 43
Benanti, Giuseppe, 89
benzenoid, 81
blind test, 27, 158–59, 161
bodega, 151, 154
Boki, Chief, 169–70
bomb, lava, 6, 65, 108, 111, 115, 139, pl40
boron, 6
Bouquet, Carole, 117
brandy, fruit, 88
breccia, 97
Bretz, J. Harlen, 164–65
Bronze Age, 11, 14
Brounstein, Al, 161
Burgundy, Duke of, 131

cactus. *See* prickly pear
Caesar, 127, 140
calcium, 1, 4, 9, 24, 75–76
caldera, 11, 14, 16, 19, 23, 30, 69, 99, 111, 117, 177n2

caper, 33, 109, 115, 120, pl43
Caravaglio, Antonio, 121
carbon dioxide, 1, 4, 37, 51
carbonic acid, 7
carbonic maceration, 152
Caro diario, 121
Cartier, Laure, 136, 142
casting of plants, victims, 45*f*, 46
Cato the Elder, 41
cellar, wine, 55*f*, 64, 128, 133–34, 140–41,
    pl47
cervoise, 127
Charlemagne, 128
Charles VII, 128
cherry, 2, 37, 50, 57, 63, 92
chestnut, 8, 45, 67–68, 77, 88, 98, 117; cask,
    79; stake, 46, 84, 90
cinder cone, 73, 77, 92, 98–100, 102, 125,
    138–39, 144, 147, 149, 154, pl51; quarry,
    138–39, 149
Circumetnea railroad, 73–78, 97, pl23
citric acid, 172
citrus fruit, 3, 50–51, 67, 75, pl29
clay, 23, 50, 55, 93, 104, 117, 119, 126, 134–35,
    142, 159, 161, 168
clementine, 75
climate, and wine, 3, 8, 22. *See also* eleva-
    tion; microclimate; orientation
clone, grape, 24, 168–69
coal, 129, 134
coffee, 2, 7–8, 169–72, pl60–61
color: of rock, soil, 9, 57; of wine, 9, 152
Columella, Lucius, 48–49
combe, 173
cone. *See* cinder cone
Conti, Giovanni, 113
Conti, Isabella, 113
Conti, Riccardo, 113
continental flood basalts, 157
contrada, 86, 90–91, 93–94, 105
Contrade dell'Etna (venue), 91
coq au vin, 128
Cremoncello, 88
Cru, Grand Cru, Premier Cru, 39, 86–87,
    90, 133, 168
Crusades, 17
crust, Earth, 4
cultivar, grape variety, 9–10, 46, 80, 149,
    154. *See also individual grape varieties*
cumulodome, 138–39
Curbelo, Don Andrés Lorenzo, 146

currais, 150
Cyclops, 98, 109, 180n1

Dahl, Johan, 54
D'Amico, Salvatore, 121
Dear Diary, 121
De Grazia, Marco, 90
Dente, Antonio, 47–48
Deroche, Jean-Baptiste, 136, 142
Dionysus, 40
DO appellation, 146, 151
DOC appellation, 37, 52, 64, 79, 90, 178n1,
    180n2
dolia, 178n8
dome, volcanic, 114, 125, 137–38
DOP appellation, 74, 179n3
Drouhin, Joseph, 166
Drouhin, Véronique, 166
Dumas, Alexandre, 51
dry waterfall, 164–65, pl57
dyke, 98, 100

earthquake, 14–15, 19, 31, 41
elevation, role of, 8, 37, 61–62, 77, 87, 92–
    94, 126, 135, 169
erosion, and soil, 6
erratic boulder, 164, 176
eruption, volcanic, 1, 15–16, 65, 68–69,
    77, 108, 144–45; eccentric, 69, 77;
    and vintage, 3. *See also individual*
    volcanoes
espumoso, 150
exposure, effect on crops, 9, 37, 49, 73, 133–
    36

feldspar, 50
Fenech, Francesco, 121
Festa del Castagno, 96
Ficodindia, 88
fig, 45, 57, 61, 109
Fiorelli, Giuseppe, 46
fissure, eruption, 51, 54, 65, 68–70, 72, 77,
    99, 146–47, 162–63
flood, catastrophic, 164, 166, pl57
flood basalts, 162. *See also* trap
foehn warming, 136
Fontana, Domenico, 38
Foti, Salvo, 84, 89–90, 94, 105, 114, pl28
Fragolino, 88
Franchetti, Andrea, 90–91, 94, 105
fruit brandy, 88

fumarole, 30, 111–12, 112*f*, 119, 139, pl34–35
furfurylthiol, 171

Gambero Rosso, 89–90
gas, 1, 14–15, 42, 54, 59, 77, 145–46; bubble, 15, 24, 115
Gauls, 127
Gault et Millau, 165
Georgas, Stefanos, 25–26
geothermal energy, 119
geyser, 160
Giordano, Angela and Ciro, 63, 17
global warming, 8
Graben, 124, 180n1
granite, 7, 130, 132, 139, 179n7
grape. *See* cultivar; *see also individual cultivars*
grapefruit, 75
gravel, 7, 159, 173–74, 179n7
Great Depression, 122
grinding stone, 40–41
Guanche, 182n1
Guénant, Bérengère, 126

Hamilton, William, 54
hang glider, 139
Hargrave, Trente, 90
Hatzidakis, Haridimos, 21
Hauner, Carlo, 109, 122
Hawaiian eruption, 4, 144, pl59
hazelnut, 67, 88, 92
heating effect of rock, soil, 9, 26, 57, 95, 149–50, 161
Henry IV, 128–29
Henry IV (Shakespeare), 145
Hephaestus, 108–9
Herodotus, 17
honey: in Hawaii, 171; on Mount Etna, 68, 92, 101
Hood, Mount, 175
hot spot, 4, 143
hoyo, 148–48, 155, pl53
Hucknall, Mick, 90

IGT appellation, 64, 110, 178n9, 179n6, 180n2
Il postino, 121
iodine, 9, 86, 110, 118, 152
iron, 2, 4, 50, 76; in Earth's core, 3

Jory soil series, 167–68, 175–76
Judgment of Paris, 158, 161, 173–74

kaolin, 114
Karamolegos, Artemis, 33
Kesteloot, Filip, 90
kouloura, 23, pl8
Koutsoyannopoulos, Georgios, 33

lactic acid, 82
Lantieri, Paola, 120, 180n6
Laurelwood soil series, 167
lava, lava flow, 2, 4, 6, 8, 40, 55, 55*f*, 62, 65, 67–72, 76–78, 91–95, 98–101, 104, 121, 123, 126, 133–35, 137, 140–41, 144, 146, 154, 157–58, 160–64, 163*f*, 166–67, 174, pl2; bomb, 6, 65, 108, 111, 115, 139, pl40, pl57, pl59; breakdown, 50, 55, 74, 116–17; diversion, 70–71; fountain, 19, 54, 65, 68–70, 77, 137, 146, 163, 179n1; lake, 138; quarry, 98, 138–39; shield, 5, 65, 69, 93–95, 125, 76; stone wall, 77, 117, 134, 148, 150, 155, pl53; tube, tunnel, 69, 72, 97, 171
lemon, 67, 76, 88, pl29
Liber Pater, 40
lichen, 50, 55
Liévain, Christian, 141, 181n14
limestone, 7, 75, 77, 93–94, 104, 125–26, 130, 132–36, 132*f*, 139–40, 159, 168, 179n7
Limoncello, 88
liquori, 88
liquoroso, 110
loam, 166–67, 173
Lo Schiavo, Nino, 109
Louis XIV, 131

macadamia nut, 171
Madonna (pop star), 117
Madonna (saint), 100, pl22
magma, 4–5, 14–15, 43, 49, 59, 78, 98, 107, 111, 115, 125, 132, 138, 143, 160, 162; chamber, 14, 16, 19, 42, 59, 145
magnesium, 2, 4
malic acid, 82, 172
malolactic fermentation, 82
mandarin, 75, 85
Mandarinetto, 85
manganese, 2
mantle, 4–6, 107, 143, 145–46, 162; plume, 49, 68, 162
marcottage, 23–24
Marie, Vincent, 139, 181n10

Marinaros, Spiridon, 17
Marino, Giovanni, 57
marl, 159
Mars, 164
Mastroberardino, 46–47
microclimate, 9, 24, 37, 61–62, 84, 95, 117,
    135–36, 160, 162, 168
mid-ocean ridge, 5, 150
mildew, 129, 145, 150
Minoan civilization, 11, 13–14, 16–17, 20,
    24, pl3–4
Moretti, Nanni, 121
Moses, parting waters of the Red Sea, 16
mud bath, 119
mudflow, 7, 54–56, 55f, 61, 64

naphthalene, 82
Nekia soil series, 167, pl58
Neruda, Pablo, 121

oak, aging in, 18, 26–27, 47, 79, 81, 86, 103,
    110, 117, 150–52
obsidian, 114, 144, 174
olive: grove, tree, 45, 50, 53, 57, 63, 67, 73,
    93, 112, 115; oil, 41, 53, 67, 126
Oliviero, Vincenzo, 55, 64
olivine, 146
onion, 149
OPAP appellation, 20, 26
orange, 3, 67, 73, 77; on Mount Etna, 67,
    73, 77, 88, 98
orientation (sun), effect on crops, 9, 37, 49,
    73, 133–36, 168
Orthodox church, 18
Ottobratta festival, 96
outlier, topographic, 134, 140, 166–67

pahoehoe, 76
palmento, 79, 85f, 90, 102–3, 105
Pappalardo, Diego, 70
Paraskevopoulos, Yiannis, 21, 32
passito, 110
PDO appellation, 20, 26–27
peach, 50, 77, 92
Peleki, Stavros, 177n6
peperite, 132–33, 132f, 140
peridotite, 146
Persilier, Gilles, 141, 181n12
PGI appellation, 28, 50, 75
Phoenician colony, 13, 17–18, 24
phonolite, 41

phosphorus, 1
phreatomagmatic eruption, 111
phylloxera, 13, 18, 23, 67, 79, 83, 109, 113–14,
    117, 129, 149
picón, 148
pillow lava, 69, 97
pine, 62, 77, 98, 100–101, 129; resin, 13
pineapple, 169, 171; wine, 171
Piromallo, Francesco, 57–58
pistachio, 3, 67–68, 74–76, 88, pl30
plate tectonics, 5, 49, 68, 107, 123, 126, 150,
    157, 159, 162, 173
Plato, 13
Plinian eruption, 5, 42, 49, 54, 59
Pliny the Elder, 39, 42, 44, 46, 48
Pliny the Younger, 5, 42–43, 45
plume. See under ash; mantle
polyphenol, 81, 84
Pomodorino del Piennolo (tomato), 51, 57–
    58, 63, pl15
Poppaea, 61
Postman, The, 121
potassium, 2, 6, 9, 24–25, 49–50, 68, 75,
    179n8
potato, 7, 147, 149
powdery mildew, 145, 150
Pradier, Jean-Pierre, 141
Pradier, Marc, 141
prickly pear, 73–74, 88
pruning, 9, 23, 84, 90, 148
pumice, 6, 9, 13–17, 21, 24–26, 25f, 30–31,
    38, 42–44, 47–48, 61, 114, 117, 136, 142,
    144, 173–74, pl5; quarry, 14, 30, 114,
    136, 142
pyroclast, 5. See also ash; cinder; scoria
pyroclastic flow, 6, 37, 43–44, 61–62, 125,
    132, 136, 138, 142, 181n5

quartz, 168

Radford, Michael, 121
rain, 3, 136
resurgent caldera, 19
rhyodacite, 24
rhyolite, 144, 174
rift, shoulder, valley, zone, 5, 73, 123–26, 132,
    134–39, 142, 180n1
riserva (oak-aged), 86, 102–3
Roman Empire, 127–28
Romeo del Castello, Rosanna, 94
rootstock, original, 18, 23, 83, 149

Roussos, Agape, 25
Roussos, Yiannis, 25
rum, 172
Russo, Domenico, 116
Russo, Maddalena, 115
Russo, Maurizio, 51–53, 62–63, pl18
Russo, Rossana, 53, 62–63

sacramental wine, 13
Sagra del Pistacchio, Bronte, 76, 96
Sagra della Fragola, Maletto, 76, 96
Sagra dell'Arancia Rossa, Centuripe, 75
salt, 24, 26, 51, 86, 110, 118, 122, 152, 180n5
sand, 7, 23, 93–94, 111, 113, 135, 139, 147–49,
    153, 174
sandstone, 7, 167
sanidine, 40
sapinière barge, 129, 134
sarcophagus, 138
schist, 7, 167
Scilio, Elizabeth, 103
Scilio, Giovanni, 103
Scilio, Luisa, 104
Sciortino, Thierry, 141
scoria, 6, 9, 49, 54, 63, 70, 77, 94, 137, 146–
    50, 154–55, 170
sherry, 150
shield volcano, 4–5, 65, 69, 76, 93–94, 125
Sigalas, Paris, 21, 24, 32, pl8
silica, 5, 24, 93, 117, 135–36, 142, 144–45,
    167–68, 174, 179n8
silicon, 1, 4, 50
silt, 165
Simply Red (band), 90
slope, and climate, 2, 9
sodium, 2, 24, 68, 117, 179n8
soil, 6–9, 24–25, 57, 93, 113, 117, 123, 148,
    157, 159–62, 166–76
sorb tree, 57, 179n14
Sorrentino, Maria Paola, 63
sparkling wine, 21, 84, 87–88, 102, 139, 141,
    150–51, 154, 171. See also spumante
spumante, 53, 63
Spurrier, Steven, 158
steam. See water vapor
steam explosion, 15–16, 30, 73, 111, 132. See
    also phreatomagmatic eruption
Stereocaulon vesuvianum, 50, 55
Stevenson, James, 112–13
stone. See heating effect of rock
Strabo, 40, 78–79

strawberry, 51, 56, 88
Strombolian eruption, 108, 115, 137, 144,
    pl40–41
sub-AVA, 161, 166–67, 169
subduction, 5, 68, 162
sub-Plinian eruption, 54
sugar: in coffee, 170; in grape juice, wine,
    6, 18, 17, 27, 50, 52, 83, 110, 117, 135, 150,
    152, 161, 169; in orange, 75; in tomato,
    51
sugarcane, 145
sulfite, 139
sulfur, 1, 9, 84, 112–13, 119, 171–72, pl35
sulfur dioxide, 1, 4
sulfuric acid, 7
syncline, 159, 173

Tacitus, Cornelius, 42
tannin, 21–22, 27, 57, 79, 81, 86, 93–95, 135,
    151, 161–62, 168
tectonic plate. See plate tectonics
terrace, agricultural, 20, 57–58, 76–77, 84,
    93, 104, 114, 117, 134–35, 140, 161
terres noires, 135
terroir, 9–10, 22, 48, 68, 82, 91, 93–94, 105,
    130, 134, 136, 140, 152, 159–61, 163, 166,
    168–73, 175–76
thermopolium, 39, pl11
Titus, 45
tomato, 2, 19, 33, 37, 51, 57–58, 147, pl15–16;
    on Mount Vesuvius, 37, 51, 57–58, pl15–
    16; in Santorini, 19, 33
trachyandesite, 139
trachybasalt, 68, 98, 179n8, 80n4
trachyte, 40–41, 138–39, 144
trap, 157, 182n1
Tre Bicchieri award, 89–90
tremor, earthquake, 14, 77, pl4
triple junction, 150
troglodyte dwelling, 142, 181n5
truffle, 126
tsunami, 16
tuff ring, 30
Turks, in Santorini, 18
Turner, William, 54

UNESCO World Heritage Site, 150

vanillin, 81
Venetian merchants, 13, 17–18, 109
Vercingetorix, 127, 140–41

vesicle, in scoria, 9
Vigo, Chiara, 94
Village appellation, 87, 127–28, 131, 135, 140, 180–81n3
Virgona, Daniela, 121
vitamin A, 75
vitamin C, 75–76
Vitruvius, 40
volatile. *See* particular gas.
volcano: active, 1; types, 3–5. *See also individual volcanoes*
Voltaire, 51
Vulcan, 109
Vulcanian eruption, 111

walnut, 57
Wars of Religion, 128
water cycle, 2
water vapor, 1, 4–5, 23, 37, 54, 69, 73, 111, 132

wheat, 41
Willakenzie soil series, 167–68
Willamette soil series, 168
Wine Olympics, 165
winepress, 33, 40, 79, 102
wine tasting: in Auvergne, 139–42; in the Canary Islands, 154–55; in Lipari, 114; on Mount Etna, 102–5; on Mount Vesuvius, 60, 62–64; in Napa Valley, 161, 172, 174; in Provence, 126; in Santorini, 31–34; in Salina, 121–22; in Willamette Valley, 175
Winiarski, Warren, 161
Woodburn soil series, 168
World Wars I and II, 129, 148
Wright, Joseph, 54

yeast, 91, 139, 141
yield, wine, 148–49

Index of Place Names:
Geographical Names,
Appellations, and Estates

References to figures are denoted by an italic "f" following the page number. "Pl" indicates numbered color plates.

All mountains are listed under their proper name rather than under the word meaning "mountain": e.g., "Etna, Mount" under *E*; "Kea, Mauna" under *K*; "Serra, Monte" under *S*.

Abona DO, 146, 151
Aci Castello, 69, 97
Acireale, 79
Aci San Antonio, 98
Aci Trezza, 97–98
Aeolian Islands, 49, 60, 106–22, 143, 145, 150, pl32; map, 107*f*
Agrigento, 80
Aix-en-Provence, 126
Akrotiri, 12*f*, 14–17, 29–31, pl3–4
Alcantara River, 77, 92, 95
Alesia, 127
Alicudi, 107, 108*f*, 114, 118
Allier River, 125, 128, 134–36, 141–42
Alps, 124
Alsace, 124
Amity, 176
Andalusia, 150–51
Andes Mountains, 5, 177n2
Antichi Vinai estate, 87–89, 104
Antiquarium, 61
Appalachian Mountains, 123
Argyros estate, 21, 25–26, 28, 34, 178n14
Arrecife, 144*f*, 154
Atlas Peak sub-AVA, 174
Atripalda, 46
Aubières, 133

Auvergne, 123–31, 136–37, 139, 141–42, pl44–49; map, 124*f*; wine cooperative, 130–31
Avernus, Lake, 51–52
Azienda Agricola Frank Cornelissen, 105, 180n16
Azienda Fuocomuorto, 55–56, 64, 179n18
Azienda Sorrentino, 62–63, 179n16
Azienda Vinicola Minardi, 117
Azores Islands, 150

Barbabecchi *contrada*, 93, pl27
Barbagallo cones, Etna, 100
Barone de Villagrande estate, 89
Beaujolais, 125, 132, 152
Beaulieu estate, volcano, 126
Benanti estate, 88–90, 92, 94, 102, 180n11
Bethel Heights Vineyard, 176, 183n10
Big Island. *See under* Hawaii
Biondi estate, 92
Black Forest Mountains, 125
Bodega El Chupadero, 155, 182n5
Bodega El Grifo, 155, 182n4
Bodega La Geria, 147*f*, 155, 182n6, pl52
Bodegas Rubicón, 155, 182n7
Bocca Nuova, Etna, 99–100
Bonjean, Stéphane, 140, 181n11, pl47

Bordeaux, comparison, cultivar, 82, 84, 86, 90, 148, 158, 165, 181
Boscoreale, 37, 50, 61
Boscotrecase, 50, 52, 54, 59, 61
Boudes, 131, 135
Boutari winery, 32, 177n6, 178n9
Bresse Graben, 124
Briare Canal, 129
Bronte, 3, 73–76, 96, pl30
Burgundy, 124–25, 129, 133, 166, 168, 174; comparison with, 134, 151, 165–67, 168–69, 173

California, 157–59, 158f, 172–74
Calistoga, 160
Calistoga sub-AVA, 174
Campania, 37, 59
Canary Islands, 7, 109, 143–55, 179, pl50
Canava Roussos, 25, 34, 178n15
Cantal, Mont du, 125, 137
Cantina del Vesuvio, 52–53, 62–63, 179n15, pl9
Cantina Valenza, 117
Cantine Olivella, 63–64, 179n17, pl17
Cape Faro, 122
Capelinhos, 150
Caravaglio, Antonio, 121
Carlo Hauner Azienda Agricola, estate, 113, 122, 180n8
Casa Barone, 56–58
Casa del Vino La Baranda, 154, 182
Cascade Range, 5, 157, 166, 174
Castellammare di Stabia, 54
Catania, 67–71, 73, 75, 77–79, 91, 96–97, 101, 119, pl19
Cave des Coteaux d'Auvergne, 130
Cave Ox pizzeria, 104
Cave Saint-Verny, 131, 141, 181n15
Cébazat, 131, 135, 140
Centuripe, 75
Chablis, 87
Chaîne des Puys, 137, pl44
Chambolle-Musigny, 166
Channeled Scablands, 163f, 164, pl57
Chanturgue, 128, 130, 135, 140
Château Beaulieu, 126
Châteaugay, 130–33, 135, 139–40
Château Montelena, 158–60, 174, 182–83n8
Chehalem Mountains sub-AVA, 166
Chiappemacine contrada, 94

Chiusa Spagnola contrada, 93
Clermont-Ferrand, 125, 127–30, 134–35, 139–40
Clos du Val estate, 159
Coast Range, 166–67, 172
Colla, Monte, 93, 179n7
Columbia River, 162, 164, 166, 176
Columbia River Basalts, 157, 163f, 164–65, 167, 175, pl57
Columbia River Highway, 175
Corent, 127–28, 130–31, 134–35, 141, pl45
Cornelissen, Frank, 91, 93–94, 105, 180n16, pl27
Corsaro, 97–98
Coteaux-d'Aix-en-Provence AOC, 126
Côtes-d'Auvergne AOC, 128, 130–32, 134–37, 140
Côtes-d'Auvergne-Boudes AOC, 135
Côtes-d'Auvergne-Chanturgue AOC, 128, 135
Côtes-d'Auvergne-Châteaugay AOC, 131–33, 132f, 135, 140
Côtes-d'Auvergne-Corent AOC, 127, 134–35
Côtes-d'Auvergne-Madargue AOC, 135
Côtes-de-Provence AOC, 126
Coudes, 136, 141
Couze Chambon, 136, 141
Couze Pavin, 142, 181n5
Couzilloux valley, 135
Crete, 13–14, 16–17, 22, 28, 109
Cristom Estate, pl58
Crown Point, 175
Cyclopes islands, 97–98

Deccan traps, 157
Delta Complex, Akrotiri, 15, 31
Diamond Mountain Creek Vineyards, 161
Diamond Mountain sub-AVA, 161
Domaine Miolanne, 142, 181n16
Domaine Sigalas. See Sigalas estate
Donnafugata, 117
Dore, Mont, 125, 137
Duckhorn Vineyards, 173
Dundee Hills sub-AVA, 166, 175

East African Rift, 169
Egypt, 13–14, 16; plagues of, 16
El Chupadero bodega, 155, 182n5
El Grifo estate, 155, 182n4
El Hierro, 143–44, 151

El Hierro DO, 146
Eola–Amity Hills sub-AVA, 166, 176, pl57
Episkopi, 25, 34
Erath Winery, 167, 175, 183n9
Ercolano, 37, 54, 55*f*, 59–60, 64
Erratic Rock State National Site, 175
Estate Argyros. *See* Argyros Estate
Esterel massif, 123
Ethiopia, 169
Ethiopian traps, 157
Etna, Mount, 2–3, 8–10, 40, 49, 60, 62, 65–105, 71*f*, 119, 125, 143; cinder cone, 73, 77, 91–92, 98–100, pl19–30; eruption of AD 252, 70; eruption of AD 1669, 70–71, 73, 77, 98; eruption of AD 1843, 73; eruption of AD 1981, 77, 95, pl25; eruption of AD 1991–1993, 72, 99–101, 103, pl21–22; eruption of AD 2002–2003, 72, 99–100; eruption of AD 2006, pl20; fest, 75–76, 96–97; fruit and nuts, 67, 74–76, 92, pl29; fruit brandies, 88; "Gold Coast," 92, 103–4; lava, 68–73, 76–78, 91–95, 98–101, 104; lava diversion, 70–73, 99; location, 66*f*, 96; orange, 67, 75, 88; Piano del Lago, 69, 99; Piano Provenzana, 101; pistachio, 74, pl30; railroad (Circumetnea), 73–78, 97, pl23; Sapienza esplanade, 68, 71, 97–100; Southeast craters, 71*f*, 99–100; summit craters, 68–69, 71*f*, 96–100; topographic map, 66*f*; Valle del Bove, 72, 87, 92, 99–100, 103; wine belt, road, 68, 78, 87, 93–94, 97, 101, 103; wine tasting, 102–5
Etna Bianco DOC, 82–83, 86
Etna Bianco Superiore DOC, 87, 92
Etna DOC, 83–84, 91
Etna Rosato DOC, 86
Etna Rosso DOC, 85*f*, 85–86
Etna Spumante DOC, 87
Exo Gonia, 33
Eyrie Vineyards, 166–67

Faial, 150
Faraglione, 112–13
Faro DOC, 80
Fay vineyard, pl54–55
Filicudi, 107, 108*f*, 109, 114, 118
Fira, 11, 12*f*, 29–30
Fiumefreddo, 101, 103–4

Forge Vecchia, 114
Foro Boario, Pompeii, 46–47, 178n8
Fossa. *See* La Fossa
Fossa delle Felci, Monte, 116, pl37
France, 18, 33, 79, 87, 93, 123, 124, 127–30, 134–35, 151, 159, 180; map, 124*f*. *See also* Auvergne; Provence
Fratelli Grasso estate, 104
Fratelli Russo brandy, 88
Fuencaliente-Las Manchas, 150
Fuerteventura, 143, 146, 153
Fuoco dell'Etna, Fuoco del Vulcano, liquor, 88
Fuocomuorto Estate, 55–56, 55*f*, 64, 179n18

Gaia wines, 21, 32, 177n5
Galfina Valley, 103
Gambino estate, 103–4, 180n15
Garajonay National Park, 153
Gavalas estate, 28, 32–33, 178n10
Gelso, 120
Gergovia, Gergovie, 127, 139–40, pl48
Gianferrante, 75
Giarre, 75, 78
Glaros Hotel, 177n6
Glass Mountain, 174
Gorna, Monte, 92, 102
Graciosa, 150
Gran Canaria, 143, 147, 150–51
Gran Canaria DO, 146
Grand Sarcoui, 138
Gravelly Meadows, 161–62
Greco, 46, 48, 83
Greece, 18, 28, 49, 109; wines, 13, 20
Güímar Valley, 151
Gurrida, Lake, 83–84

Haleakala, 171
Hadzidakis winery, 22, 28, 32, 177n7
Hauner estate, 113, 122, 180n8
Hawaii, 4, 153, 169–71; Big Island, 153, 169–71, pl59–60
Hennessey, Lake, 173
Herculaneum, 2, 6, 35–37, 36*f*, 39, 43–44, 54, 60–61, 64, pl11
Hercynian Mountains, 123
Hermitage, 93, 179n7
Highway 29, 160, 172
Hood, Mount, 175
Hotel Conti, 113

Hotel Corsaro, 97–98
Howell Mountain sub-AVA, 174
Hualalai, 169–70

Idaho, 157, 162–64
Il Cantante estate, 90
Indonesia, 5, 7, 16
Inglenook estate, 158
Ischia, 49
Isole Ciclopi, 97–98
Italy, 2,5, 21, 37–38, 49–50, 67, 74–75, 79–
    80, 83, 88, 109, 124, 177
I Vigneri, 79, 84, 85f, 88–90, 94, 105, 114,
    180n17

Joan of Arc, 128
Jura Mountains, 151

Kaiserstuhl, 124
Kalliste, 17
Kamari, 29, 32
Kameni, 19, 30. *See also* Nea Kameni
Karamolegos estate, 28, 33, 178n13
Kea, Mauna, 154, 169
Kilauea, 71, 169, 171
Kona, 8, 170–71, pl60–61
Konocti volcano, 162
Knossos, 14

La Alborada, 154
La Florida, 154
La Fossa, 111, 112f, 113, 119, pl33–35
La Fruttiera, 95, pl25
La Geria, 144f, 148
La Geria estate, 147f, 155, 182n6, pl52
La Gomera, 143, 151–53
La Gomera DO, 146
Lajido, 150
Lanai, 169
Lanzarote, 143–44, 144f, 146–51, 153–55,
    pl50–53; eruption of AD 1730–1736,
    146–47, 154–55, pl51
Lanzarote DO, 146
La Palma, 143–44, 146, 150–51
La Palma DO, 146
Latourell Falls, 175
Le Case di Civi, 75
Leilani Estates, pl59
Lemptégy, 138–39, 181n7
Limagne Graben, 123–26, 132, 134–37, 139,
    pl46

Limagrain, 131
Lingua, pl37
Linguaglossa, 78, 90, 92, 103–4
Lipari, 60, 107–9, 108f, 113–14, 116, 118–19;
    wine tasting in, 114
Liquori Fichera, 88
Loa, Mauna, 5, 8, 154, 169–71
Loing River, 128
Loire River, 128–29
Los Bermejos estate, 155

Madargue, 130, 135
Madeira, 109, 145, 151–52
Maletto, 76, 96
Malfa, 117, 121–22
Malvasia delle Lipari DOC, 109–10, 114,
    116–17
Malvasia Dolce, 152
Malvasia Seco, 151
Marmajuelo, 150
Mars, 164
Marsala, 83
Martes-de-Veyre, 134, 141
Mascali, 75, 78, 80
Massa, 56
Maui, 171
Maui Wine estate, 171
Mayacamas Mountains, 159–61
McMinnville, 166, 176
Megalochori, 32–33
Ménétrol, 140
Meso Gonia, 34
Mesopotamia, 13
Messina, 80
Milazzo, 60, 108f, 111, 114, 119
Milo, 82, 87, 92
Miolanne estate, 136, 142
Misena, Cape, 42
Mississippi River, 165
Missoula, Lake, 176
Missoula Floods, 163f, 164, 166–67, 176,
    pl57
Misterbianco, 73
Monemvasia, 109
Montana, 165–66
Montelaguardia, 93, 95
Morgex, 179n4
Morgon, 86
Multnomah Falls, 175
Mumm Napa Winery, 173
Museum of Prehistoric Thira, 30

Napa (city), 173
Napa Valley, 148, 157–62, 158*f*, 166, 172–75, 173*f*, pl54–56
Naples, 36*f*, 37–38, 47–48, 50, 52–54, 60–61, 63, 119; Bay of, 35, 36*f*, 38, 42, 51–52, 57, 60–61, 64
Naxos, 17
Nea Kameni, 12*f*, 19, 30, pl2
Neschers, 136, 141
Newberg, 175
Niagara Falls, 165
Nicolosi, 70, 97–98
Nicosia Azienda winery, 92, 101–2, 179–80n10
Nikola's Tavern, 31
Nocera, 38, 113
No Control estate, 181n10
Nola, 38
North Atlantic traps, 157
Nuestra Señora de la Caridad, 155
Nykteri, 26–27, 32–34, pl7
Nyos, Lake, 179n13

Oahu, 169–70
Oia, 11, 12*f*, 24, 29–32, pl1
Old Faithful, 160, 174
Oplontis, 61
Orcines, 138
Oregon, 3, 5, 9, 157–58, 158*f*, 162–69, 172, 175
Orvietto, 41
Osservatorio pizzeria, 120–21
Oste Eusinio, Pompeii, 47
Ottaviano, 54

Pacific Coast, Pacific Ocean, 5, 159, 162–66, 169
Pacific Highway West, 175
Palermo, 80, 96, 117–18
Palomino Fino, 150
Panarea, 107, 108*f*, 118
Pantelleria, 117
Parco Nazionale del Vesuvio, 62
Paris, 128–29, 134, 158, 165; Judgment of, 158, 161, 172, 174
Passito di Pantelleria, 117
Passopisciaro, 78, 82, 88, 91, 93, 104–5, pl26
Passopisciaro estate (Vini Franchetti), 105
Paternò, 70, 73, 75
Pedara, 98

Perrier, 142, 181n5
Petto Dragone *contrada*, 104
Phlegrean Fields, 49, 51
Piano del Lago, Mount Etna, 69, 99
Piano Provenzana, 101
Pico, 150
Piedimonte Etneo, 92, 103
Pilato, Monte, 108
Pompeiano IGT, 47
Pompeii, 2, 6, 15–16, 35–50, 36*f*, 45*f*, 54, 59–62, 60*f*, pl10, 12; amphitheater district, 39, 46, 61; vineyards today, 46–47, 61
Pompignat, 131, 140
Pontgibaud, 138
Porcaria *contrada*, 94
Porri, Monte dei, pl38
Portici, 54
Portland, 174
Portugal, 109, 145, 151, 177, 182
Provence, 126
Puy Chopine, 138
Puy de Côme, 138–39, pl44
Puy-de-Dôme (department), 129–30
Puy de Dôme (volcano), 138–39
Puy de Lassolas, 137
Puy de la Vache, 137
Puy de Pariou, 138–39, pl44
Puy de Sancy, 125, 136–37, 142
Pyrenees Mountains, 124
Pyrgos, 25, 32–33

Quattropani, 114

Rampante *contrada*, 93–94
Randazzo, 76–78, 90, 93, 95
Red Rock Terrace, 161
Rhine Graben, 124
Rhodes, 21–22
Ribbon Ridge, 166
Rifugio Ragabo, 97, 101
Rifugio Sapienza, 97–98
Riom, 130–31, 140
Riposto, 78–79, 88
Roche Rosse, 114
Rognes, 126
Rolle, 126
Roman Comagmatic Province, 49
Rombauer Vineyards, 174, 182n7
Rome, 41, 127–28
Romeo del Castello estate, 94, pl25
Rossi, Monti, 70, 98

Roussos estate, 25, 34, 178n15
Rovitello, 78, 104
Rutherford, 173

Saint Helena, Mount, 160, 174
Saint Helena Highway, 174
Saint Helena sub-AVA, 173–74
Saint-Nectaire, 142
Saint-Pierre, Martinique, 6
Saint-Pierre-Colamine, 142
Saint-Verny, 131, 141, 181n15
Salerno, 39, 178
Salina, 107–10, 108*f*, 113, 116–19, 121–22,
    180n5, pl37–39
Salina IGT, 110, 118
San Andreas Fault, 157
San Bartolomé, 154
San Francisco, 158–59, 172
San Giuseppe Vesuviano, 54
San Sebastiano, 50, 56, 63
Santa Marina, 121–22
Sant'Anastasia, 63
Santa Venerina, 87–88, 103
Santo Nicola, Monte, 92
Santorini (appellation), 26, 32–34
Santorini (island), 9–34, 12*f*, 49, 69, 148–
    49, pl1–8
Santo Wines, 19–20, 22, 28, 31–33, 178n11
Sapienza esplanade, Mount Etna, 68, 71,
    97–100
Sarno, 7
Sarno River, 38, 42
Scablands. *See* Channeled Scablands
Scamacca del Murgo estate, 87, 89, 103,
    180n12
Schiena dell'Asino trail, 100
Sciara del Fuoco, 115, 121
Sciara Nuova *contrada*, 94
Scilio estate, 89, 103–4, 180n14
Sciortino, Thierry, 141
Seine River, 128–29
Serra, Monte, 92, 98, 102
Serra della Contessa, 92, 102
Sicilia IGT, 84, 110
Sicily, 3, 8, 40, 48–49, 60–61, 67–68, 74–
    75, 80–81, 83–84, 86, 88–89, 92, 96,
    102, 104–5, 107, 111, 114, 117, 119, 145,
    150, pl32; map, 66*f*, 83, 108*f*
Sigalas estate, 21, 25, 28, 32, 178n8
Silverado Trail, 160, 173–74
Silvestri, Monti, 99

Simeto River, 75
Solicchiata, 78, 93, 104
Somma, Monte (paleo-Vesuvius), 49, 56, 62
Sonoma Valley, 159–60
Sorrentino estate, 62–63, 179n16
Sorrento Peninsula, 47
Spain, 124, 145, 149–50, 152, 109, 182
Stabiae, 35, 36*f*, 44
Stag's Leap Wine Cellars, 158–59, 161, 173,
    182n6, pl54–55
Stags Leap sub-AVA, 160, 173, 182n3
St. Helena, Mount, 160, 174
St. Helens, Mount, 3, 136, 157, 160
Stromboli, 60, 107–8, 108*f*, 110, 114–18,
    120–21, pl31, 40–42; eruption of
    AD 1930, 115–16; Sciara del Fuoco, 115,
    121
Strongyle, 17
Suez Canal, 14
Sunset Crater, 125

Tacoronte-Acentejo DO, 146, 151
Tambora, 16
Tanzania, 8
Tasca d'Almerita, 122, 180n7
Teide, Mount, 146, 153–54
Tenerife, 143, 145–46, 150–54
Tenuta Capofaro, 180n7
Tenuta delle Terre Nere estate, 90
Tenuta di Castellaro estate, 114, 180n3
Tenuta San Michele B&B, 103, 180n13
Tenuta Scilio estate, 89
Terra & Sorsi, 102
Terra di Trente estate, 90
Thira, 17–18, 21
Timanfaya National Park, 144*f*, 146, 154,
    pl51
Torre Annunciata, 61
Torre del Greco, 54
Tournoël castle, pl46
Trapani, 117
Trecase, 37, 59, 62
Trecastagni, 79, 81, 91, 97, 101
Troina, Monte, 98
Tuscany, 81, 83, 90

United States of America, 157, 160, 169

Vaca Mountains, 161–62, 173, pl54
Valdichiesa, 116–17, 122, pl38
Valle de Güímar DO, 146, 151

Valle de la Orotava DO, 146
Valle del Bove, Etna, 72, 87, 92, 99–100, 103
Valle dell'Inferno, Vesuvius, 62
Valle Galfina *contrada*, 103–4
Van Duzer Corridor, 166–67
Verdelho de Pico, 150
Vesuvio DOC, 37, 51
Vesuvius, Mount, 2, 5, 9, 16, 35–45, 36*f*, 47–
    51, 54, 55*f*, 56, 58–62, 58*f*, 60*f*, 69, 108,
    119, 125, 143, pl9–18; crater, 62; erup-
    tion of AD 79, 35, 41–45, 49, 59; erup-
    tion of AD 1631, 35, 49, 55, 59, 62;
    eruption of AD 1906, 49, 53–57, 62,
    64; eruption of AD 1944, 35, 56–57,
    58*f*, 59, 62, pl16; fruit and vegetables,
    50–51, 57–58; magma composition, 49;
    National Park, 62; Somma, Monte, 49,
    56, 62; tectonic setting, 49; tomato,
    51, 57–58; Valle dell'Inferno, 62; wine
    belt, 51
Veyre-Monton, 141
Viagrande, 79, 81–82, 88, 91, 98, 102
Vic-le-Comte, 141
Villa Baulieu, 126
Villa dei Misteri (Villa of the Mysteries),
    40; wine, 47–48
Vini Franchetti, 105, 180n18
Vinsanto, 11, 18–19, 26–28, 33–34, 177n1,
    pl8

Visperterminen, 179n4
Volcan de Lemptégy, 138–39, 181n7
Volcanic Hill, 161–62
Volcano Winery, 171
Volcan Wines, 33, 178n12
Volvic, 139, 181n9
Voragine, Etna, 99–100
Vosges Mountains, 125
Vothonas, 33
Vulcanello, 111, 113
Vulcania, 138, 181n6
Vulcano, 60, 107–13, 112*f*, 117–19, 118*f*,
    pl33–36; eruption of AD 1888–1890,
    108, 111; La Fossa, 111, 112*f*, 113, 119–20,
    pl34–36; Vulcano Piano, 111, 113, 119–
    20

Washington State, 3, 5, 157, 162, 164
Willamette Valley, 9, 157–58, 158*f*, 162–63,
    165–69, 171, 175–76, pl58

Xeres, 150

Yaiza, 146
Yamhill-Carlton District, 166
Ycoden-Daute-Isora DO, 146

Zafferana Etnea, 72, 92, 96–97, 100–103,
    pl21–22

Index of Grape Cultivars

References to figures are denoted by an italic "f" following the page number; "pl" denotes numbered color plates.

Aglianico, 46–48, 52, 56, 63
Aidani, 18, 21, 25–28, 32–34
Albillo, 152
Alicante, 83, 113
Assyrtiko, 10, 13, 18, 20–21, 24, 25*f*, 26–27, 32, pl6–8
Athiri, 18, 21, 26–28
Auvernat, 130

Black Corinth, 110, 113–14, 116, 118

Cabernet Franc, 90
Cabernet Sauvignon, 82, 84, 126, 150, 158–59, 161–62, 171, 174, pl56
Caprettone, 46, 48, 52, 56, 58, 64, pl13
Carignan, 171
Carnelian, 171
Carricante, 68, 82–83, 86–88, 91–92, 102, 114
Catalanesca, 64
Catarratto, 83, 86, 117–18
Cayuga White, 171
Chambourcin, 171
Chardonnay, 84, 88, 105, 130–31, 133, 139, 158–59, 174
Chenin Blanc, 171
Coda di Volpe, 46, 48, 52–53, 83
Corinto Nero. *See* Black Corinth

Diego, 150, 155
Dijon clone. *See under* Pinot Noir

Falanghina, 46, 48, 52–53
Fiano, 46, 51
Forstera Blanca, 150

Gamay, 129–32, 135, 139, pl49; d'Auvergne, 130, 141
Greco, 46, 48, 83
Grenache, 83–84, 113, 126, 171
Grillo, 118
Gual, 150

Inzolia, 83, 104, 117
Isabella, 150

Katsano, 28, 33

Lajido, 150
Listán Bianco, 149–52
Listán Negro, 150–52, 182n2

Malbec, 171
Malvasia, 107, 109–10, 113–14, 116–19, 121, 145, 149–52, 155, pl36, pl39
Mandilaria, 22, 28, 32, 34
Maréchal Foch, 171
Marmajuelo, 150
Mavrotragano, 21–22, 28, 32–34
Merlot, 81–82, 84, 150, 174
Minnella, 82–83, 85–86
Minutidda. *See* Black Corinth

Mission Grape. *See* Listán Negro
Muscat, Moscatel, 83, 117, 149, 152, 155, 171

Nebbiolo, 21
Negramoll, 151–52
Nerello Cappuccio, 81–82, 85–86, 88
Nerello Mantellato. *See* Nerello Cappuccio
Nerello Mascalese, 10, 68, 80–82, 85*f*, 85–91, 95, 105, 116, 118, pl24, pl28
Nero d'Avola, 80, 83, 88, 104, 113–14, 116–18
Nireddu. *See* Nerello Mascalese
Nocera, 113

Olivella. *See* Sciascinoso

Palomino Fino. *See* Listán Bianco
Pellechiella, 50
Petit Verdot, 90, 105
Piedirosso, 9, 46–48, 52–53, 56, pl14
Pinot Grigio, 171
Pinot Noir, 3, 81, 84, 129–31, 133, 135–36, 139, 141, 165–69, 171, 175–76; Pommard and Dijon clones, 168

Pommard clone. *See under* Pinot Noir
Primitivo, 56

Riesling, 82
Rolle, 126

Sangiovese, 81, 113
Sciascinoso, 46–48, 52
Sémillon, 126
Syrah, 84, 116, 126, 130

Tintilla, 151
Trebbiano, 83, 85–86
Trousseau, 151

Verdelho, Verdello, 88, 150
Vesparola, 83
Vijariego. *See* Diego
Viognier, 171
Voudomato, 22, 28, 32–33

Xenoloo, 28

Zibbibo. *See* Muscat, Moscatel
Zinfandel, 174